APPLIED GEOPHYSICS IN PERIGLACIAL ENVIRONMENTS

Many important problems in cryospheric science, such as global-warming-induced permafrost degradation, concern subsurface properties and processes that take place some metres below the surface. Geophysical techniques can be used to study ground ice and characterise areas of permanently frozen ground, but surveys in mountainous and polar areas demand specialised techniques for sensor coupling; data acquisition and interpretation; inversion routines; and logistical issues in cold and remote environments.

This book starts with an introduction to the main geophysical methods and then demonstrates their application through case studies written by a team of international experts in the various field techniques. The final part of the book presents a series of reference tables with typical values of geophysical parameters for periglacial environments.

Written as a reference guide for the application of geophysical techniques in mountainous and polar terrain, this will serve as a handbook for planning and conducting field surveys. It is a valuable resource for glaciologists, geomorphologists and geologists requiring an introduction to geophysical techniques, as well as for geophysicists lacking experience of working in periglacial and glacial environments.

CHRISTIAN HAUCK received a Ph.D. from the Eidgenössische Technische Hochschule (ETH), Zürich. He has been a scientist at the Institute for Meteorology and Climate Research at the Karlsruhe Institute of Technology for 7 years, and is now a Professor at the University of Fribourg, Switzerland. He has 10 years of experience in conducting field work in alpine, arctic and antarctic environments using electrical, electromagnetic and seismic surveying methods. Dr Hauck's main research interests include the application of geophysical methods in mountainous and polar terrain, and new measuring and inversion methods in hydrogeophysics.

CHRISTOF KNEISEL received a Ph.D. from the University of Trier. He is now a Lecturer in the Department of Physical Geography at the University of Würzburg where he finished his professorial dissertation in 2007. He has 12 years of experience in conducting field work in alpine and subarctic periglacial environments using electrical and seismic surveying methods. Dr Kneisel's main research interests include the application of geophysical mapping and monitoring in alpine and subarctic periglacial terrain on various glacial and periglacial landforms, and the application of geoelectrical methods in soil and environmental science.

APPLIED GEOPHYSICS IN PERIGLACIAL ENVIRONMENTS

C. HAUCK

University of Fribourg, Switzerland

and

C. KNEISEL

University of Wüzburg, Germany

CAMBRIDGE UNIVERSITY PRESS
Cambridge, New York, Melbourne, Madrid, Cape Town,
Singapore, São Paulo, Delhi, Mexico City

Cambridge University Press
The Edinburgh Building, Cambridge CB2 8RU, UK

Published in the United States of America by Cambridge University Press, New York

www.cambridge.org
Information on this title: www.cambridge.org/9781107406193

First published 2008
First paperback edition 2012

A catalogue record for this publication is available from the British Library

Library of Congress Cataloguing in Publication Data
Hauck, C.
Applied geophysics in periglacial environments/C. Hauck, C. Kneisel.
p. cm.
Includes bibliographical references and index.
ISBN 978-0-521-88966-7 (hardback)
1. Geophysics–Data processing. 2. Periglacial processes.
I. Kneisel, C. II. Title.
QE501.3.H38 2008
551.31–dc22   2008016679

ISBN 978-0-521-88966-7 Hardback
ISBN 978-1-107-40619-3 Paperback

Contents

The colour plates are situated between pages 80 and 81, and are also available for download from www.cambridge.org/9781107406193

Contributors

Halfdan Benjaminsen Norwegian Water Resources and Energy Directorate, PO Box 5091 Majorstua, N-0301 Oslo, Norway.

Ivar Berthling Department of Geography, Norwegian University of Science and Technology, Dragvoll, N-7491 Trondheim, Norway.

Reynald Delaloye Department of Geosciences, University of Fribourg, Pérolles, CH-1700 Fribourg, Switzerland.

Wojciech Dobinski Department of Geomorphology, University of Silesia, ul. Bedzinska 60, 41-200 Sosnowiec, Poland.

Bernd Etzelmüller Department of Geosciences, University of Oslo, PO Box 1047 Blindern, N-0316 Oslo, Norway.

Herman Farbrot Norwegian Meteorological Institute, PO Box 43 Blindern, N-0313 Oslo, Norway.

Koichiro Harada Department of Environmental Sciences, School of Food, Agricultural and Environmental Sciences, Miyagi University, Hatatate 2-2-1, Taihaku-ku, Sendai 982-0215, Japan.

Christian Hauck (Editor) Institute for Meteorology and Climate Research, Forschungszentrum Karlsruhe, PO Box 3640, D-76021 Karlsruhe, Germany.

Andreas Hördt Institute for Geophysics and Extraterrestrial Physics, TU Braunschweig, Mendelssohnstrasse 3, D-38106 Braunschweig, Germany.

Thomas Hoffmann Geographical Institute, University of Bonn, PO Box 1147, D-53001 Bonn, Germany.

Atsushi Ikeda Graduate School of Life and Environmental Sciences, University of Tsukuba, Tsukuba, Ibaraki 305-8572, Japan.

Tristram D. L. Irvine-Fynn Department of Geography, University of Sheffield, Winter Street, Sheffield, S10 2TN, UK.

Ketil Isaksen Norwegian Meteorological Institute, Climatology Division, PO Box 43 Blindern, N-0313 Oslo, Norway.

Mamoru Ishikawa Faculty of Environmental Earth Science, Hokkaido University, Kita-10 Nishi-5 Kita-ku, 060-0810 Sapporo, Japan.

Christof Kneisel (Editor) Department of Physical Geography, University of Würzburg, Am Hubland, D-97074 Würzburg, Germany.

Ånund Kvambekk Norwegian Water Resources and Energy Directorate, PO Box 5091 Majorstua, N-0301 Oslo, Norway.

Christophe Lambiel Institute of Geography, University of Lausanne, Quartier Dorigny, Bâtiment Anthropole, CH-1015 Lausanne, Switzerland.

Leszek Litwin Institute ISPiK S.A., ul. Dworcowa 56, 44-100 Gliwice, Poland.

Kjetil Melvold Norwegian Water Resources and Energy Directorate, PO Box 5091 Majorstua, N-0301 Oslo, Norway.

Brian J. Moorman Department of Geography, University of Calgary, 2500 University Drive NW, Calgary, AB, T2N 1N4, Canada.

Rune Ødegård Faculty of Engineering, Gjøvik University College, PO Box 191, N-2802 Gjøvik, Norway.

Oliver Sass Institute of Geography, University of Augsburg, Universitätsstrasse 10, D-86135 Augsburg, Germany.

Lothar Schrott Department of Geography and Geology, University of Salzburg, Hellbrunnerstrasse 34, A-5020 Salzburg, Austria.

Thomas V. Schuler Department of Geosciences, University of Oslo, PO Box 1047 Blindern, N-0316 Oslo, Norway.

Morgan Wåle Geophysik AS, Nesle, N-3692 Sauland, Norway.

Krystian Wzietek Department of Applied Geology, University of Silesia, ul. Bedzinska 60, 41-200 Sosnowiec, Poland.

Bogdon Zogala Department of Applied Geology, University of Silesia, ul. Bedzinska 60, 41-200 Sosnowiec, Poland.

Introduction

Climate warming and its impact on periglacial environments is a research topic of increasing importance, due to the growing concern of warming-induced permafrost degradation and its consequences regarding slope instabilities, construction failure and other hazards related to the melting of ground ice. Periglacial environments are regions with cold, and generally non-glacial conditions, in which frost-related processes and/or permafrost are either dominant or characteristic (French 2007).

Periglacial environments can be highly variable regarding surface and subsurface conditions. The ground thermal regime (e.g. the occurrence of permafrost) depends strongly on altitude, incoming radiation, local climatic conditions and surface and subsurface factors (e.g. organic layers, characteristics of unconsolidated sediments such as coarse blocky material). These often heterogeneous surface and subsurface conditions call for methods that are able to resolve the shallow subsurface at scales between a few metres and several kilometres. Due to the high costs and the logistical difficulties, deep borehole information is scarce and can only be used as point information at special sites. In contrast, geophysical methods are able to characterise the subsurface continuously over larger areas, often with investigation depths down to 100 metres, and can be applied with comparatively little financial and personal effort. Taking into consideration the high costs of drilling, such desirable ground-truthing is seldom possible and can be seen as one of the main reasons for the application of geophysical methods.

Geophysical methods are used to gain information about the physical properties and the structure of the subsurface and have been used for many years to study ground ice and characterise areas of permanently frozen ground. Until the late 1980s they were mostly applied in polar permafrost regions, where seismic, electromagnetic and electrical methods were particularly suitable for exploration and engineering purposes. More recently a number of geophysical applications in mountain regions using state-of-the-art tomographic measurement systems have been published. In spite of the recent successes, the application of geophysical methods in periglacial environments is not as straightforward as their counterparts on non-frozen, homogeneous and logistically easier to handle terrain, such as vegetated or non-vegetated soils and hill slopes. Geophysical surveys in mountainous and polar areas demand specialised techniques for sensor coupling, data acquisition and interpretation, inversion routines, and logistical issues in cold and remote environments (e.g. power supply, transport, safety issues).

This book has been composed as a reference for the application of geophysical techniques in mountainous and polar terrain and should serve as a handbook for planning and conducting field surveys. It is aimed at graduate and postgraduate students as well as senior researchers from all disciplines of the earth sciences, such as glaciologists, geomorphologists, hydrologists and geologists without special knowledge of geophysical techniques, as well as geophysicists without the experience of working in periglacial and glacial environments.

Part I introduces the most commonly applied techniques in periglacial research in the form of textbook chapters with special emphasis on the application and particularities of periglacial terrain. At the end of each of the four chapters (describing electrical, electromagnetic, seismic and GPR techniques) a short checklist for survey preparation, data acquisition and data processing is presented. A comprehensive literature review is given in each chapter. This part focuses on the most commonly applied methods excluding borehole geophysical techniques, since such techniques require the existence of boreholes, which are often difficult to obtain in periglacial terrain.

In Part II several international scientists present specific applications of geophysical techniques in the form of short case studies, including surveys on rock glaciers, moraines, glacier forefields, talus slopes, bedrock, glaciers, snow and river ice, and covering geographic locations reaching from the high Arctic (including Svalbard, Canada and Alaska), over the central Asian mountain ranges (Himalayas and Mongolia) to Japan, as well as to the European Alps (Switzerland, Austria, Italy and Germany) and the central European highlands (including the Polish Tatra, the Black Forest and the Vosges Mountains). The case studies are written as short examples of the methods introduced in Part I, and are often summaries of published studies, which are referenced for further reading.

The most important findings are summarised in the Appendix in the form of comprehensive tables. The tables should aid the periglacial researcher in choosing the best method for a specific problem, and provide baseline material properties of common subsurface materials for data interpretation. Additionally, the characteristics of each method (such as penetration depth, power requirements, persons needed for surveying, and data processing) are presented.

REFERENCE

French, H. (2007). *The Periglacial Environment*. John Wiley & Sons.

Part I

Geophysical methods

1

Electrical methods

C. Kneisel and C. Hauck

1.1 Introduction

Commonly used electrical methods in applied geophysics comprise direct-current
(DC) electrical measurements, self-potential measurements (SP) and induced-
polarisation methods (IP), including spectral-induced polarisation (SIP).

The SP method is based on passive measurements of natural electrical
potential differences in the ground, which are often negligible in periglacial areas
as electrically conductive materials or water flow have to be present to generate
distinct SP patterns. A cryospheric example measuring subglacial drainage
conditions with SP is given by Kulessa *et al.* (2003).

The IP and SIP methods are based on actively induced polarisation effects in the
subsurface, which require polarisable material to be present. Again, these effects
are usually small in frozen environments, which is why all three methods have
seldom been used in periglacial research to date. A review concerning SP and IP
methods is included in the review paper concerning the application of geophysical
methods in permafrost areas by Scott *et al.* (1990). Further details on these techniques are given in Weller and Börner (1996) and Slater and Lesmes (2002).

In contrast, DC resistivity methods utilise distinct changes in the electrical
resistivity within the subsurface, and constitute one of the traditional geophysical
methods that have been applied in permafrost research. Since a marked increase
of the electrical resistivity occurs at the freezing point, these methods are
expected to be most suitable to detect, localise and characterise structures con-
taining frozen material. Based on the number of scientific publications in the past
decade and the large variety of applications, the tomographic variant of the
method (electrical resistivity tomography, ERT) is maybe the most universally
applicable method for research in periglacial permafrost-related mountain

Applied Geophysics in Periglacial Environments, eds. C. Hauck and C. Kneisel. Published by Cambridge
University Press. © Cambridge University Press 2008.

environments (in combination with another geophysical method, if possible, e.g. Hauck and Vonder Mühll 2003a, Kneisel and Hauck 2003). Due to the recent development of multi-electrode resistivity systems and commercially available two-dimensional inversion schemes for data processing, this method is comparatively easy to apply even in very heterogeneous mountain and arctic terrain. As for most geophysical techniques, the obtained resistivity model is not unambiguous and depends strongly on data quality, measurement geometry and the choice of inversion parameters.

In this chapter the measurement principles of DC resistivity soundings (also called vertical electrical soundings, VES) and ERT are introduced, including data acquisition and processing as well as a discussion of various pitfalls in resistivity inversion and interpretation concerning the detection and characterisation of subsurface materials in periglacial environments.

1.2 Measurement principles

Resistivity surveys are conducted by injecting a direct electrical current (I) into the ground via two current electrodes (A and B in Figure 1.1). The resulting voltage difference (ΔV) is measured at two potential electrodes (M and N). The overall purpose of resistivity measurements is to determine the subsurface resistivity distribution. From the current (I) and voltage difference values (ΔV) the resistivity ρ is calculated using

$$\rho_a = K \frac{\Delta V}{I},\qquad(1.1)$$

where K is a geometric factor that depends on the arrangement of the four electrodes. This calculated resistivity value is not the 'true' resistivity of the subsurface, but a so-called 'apparent resistivity' ρ_a, which equals the 'true' (or specific) resistivity only for a homogeneous subsurface. For heterogeneous

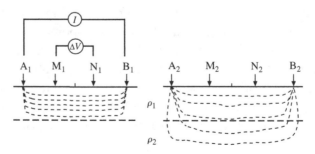

Figure 1.1.   Conventional four-electrode configuration in resistivity surveys.

Table 1.1. *Range of resistivities for different materials*

Material	Range of resistivity (Ω m)
Clay	1–100
Sand	$100-5 \times 10^3$
Gravel	$100-4 \times 10^2$
Granite	$5 \times 10^3-10^6$
Gneiss	$100-10^3$
Schist	$100-10^4$
Groundwater	10–300
Frozen sediments, ground ice, mountain permafrost[a]	$1 \times 10^3-10^6$
Glacier ice (temperate)	10^6-10^8
Air	infinity

[a] Kneisel (1999)
Compiled mainly after Telford *et al.* (1990) and Reynolds (1997)

resistivity distributions in the ground the resistivity can be derived from the measured apparent resistivity values using inversion methods implemented, for instance, in commercially available software programmes (see below).

The basic principle for the successful application of geoelectrical methods in geomorphology/quaternary geology is based on the varying electrical conductivity (= 1/resistivity) of minerals, solid bedrock, sediments, air and water, and consequently their varying electrical resistivity (Table 1.1). Resistivity surveys give an image of the subsurface resistivity distribution. Knowing the resistivities of different material types, it is possible to convert the resistivity image into an image of the subsurface consisting of different materials. However, as a consequence of overlapping resistivity values of different materials, this conversion might be non-unique. The resistivity of rock, for example, depends on water saturation, chemical properties of pore water, structure of pore volume and temperature. The large range of resistivity values for most materials is thereby due to varying water content. Resistivity values for frozen ground can vary over a wide range (from between 1 and 5 kΩ m to several hundred kΩ m or even a few MΩ m: e.g. Hoekstra and McNeill 1973, Haeberli and Vonder Mühll 1996, Kneisel 1999, Ishikawa *et al.* 2001, Hauck and Vonder Mühll 2003a, Marescot *et al.* 2003, Kneisel 2006, Kneisel *et al.* 2007). Apart from the host material (lithology and textural characteristics of the frozen material), the resistivity depends on the ice content, the temperature, and the content of impurities. The dependence of resistivity on temperature is closely related to the unfrozen water content; as in most earth materials, electrical conduction takes place through ionic transport in the liquid phase.

When the distance between the current electrodes (A, B) is increased, a larger penetration depth is obtained (as indicated in Figure 1.1) yielding more information about the deeper sections of the subsurface. The penetration depth d depends on the measurement geometry and is limited by the maximum electrode spacing. It may be estimated using a formula by Barker (1989) with $d = 0.17L$ for the so-called Wenner array (see below), with L being the distance between the outer (current) electrodes.

1.2.1 Measurement configuration and array types

Vertical electrical soundings (VES)

For electrical resistivity surveys different array types are used. In traditional one-dimensional DC resistivity soundings (or vertical electrical soundings, VES) the symmetrical Schlumberger array is applied (see Chapters 5, 7, 8 and 9). In Schlumberger surveys the distance between the outer current electrodes is increased logarithmically to obtain a sounding curve with maximum penetration depth, while the distance between the potential electrodes remains mostly constant. For the interpretation of one-dimensional data the assumption is made that the subsurface consists of horizontal layers and that the resistivity changes only with depth but not horizontally. The obtained resistivity values are interpreted as a one-dimensional layered model of the subsurface using standard software packages.

Wenner profiling

As another classical survey technique, the resistivity profiling method is used for obtaining lateral changes in the subsurface resistivity. In this case, the Wenner array is applied where the spacing between the electrodes remains fixed, but all four electrodes are moved simultaneously for each reading. Wenner profiling is used to obtain information about lateral changes, but not about vertical changes in resistivity.

Electrical resistivity tomography (ERT)

From the description above, the limitations of Schlumberger sounding and Wenner profiling surveys are evident. The assumed horizontal layering as well as the assumption that resistivity changes only with depth but not horizontally will not always be valid in practice. On heterogeneous ground conditions, the interpretation of one-dimensional soundings can be difficult, as lateral variations along the survey line can influence the results significantly. The sounding curve produces an average resistivity model of the survey area. For some studies this might not be problematic and the results obtained from one-dimensional soundings are sufficient. However, individual anomalies will not show explicitly

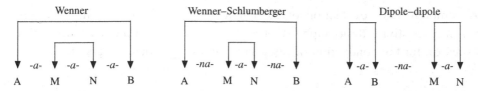

Figure 1.2.   Schematics of the most commonly used array geometries in ERT surveys.

in the results. Two-dimensional resistivity tomography (ERT) overcomes this problem using multi-electrode systems and two-dimensional data inversion yielding a more accurate model of the subsurface (see Chapters 6, 8, 9 and 10). ERT requires multiple resistivity measurements with various electrode spacings along a profile line (2D) or on a two-dimensional grid (3D). The most commonly used measurement geometries in ERT surveys are the Wenner, Wenner–Schlumberger and Dipole–dipole arrays (see Figure 1.2).

Array types

In Wenner surveys, the two outermost electrodes (A and B) are used as current electrodes while the potential difference is measured at two electrodes in between (M, N). Potential-electrode spacing increases as current-electrode spacing increases, with equal distances between all electrodes for each measurement. The Wenner configuration has a moderate investigation depth and good resolution for horizontal structures that change with depth. Since the total number of measurements required is smaller than for other configurations, the time to complete a survey is comparatively short; however, less information for the subsurface is obtained than from other arrays. The Wenner–Schlumberger array is a combination of the Wenner array and the Schlumberger array with constant potential-electrode spacing but increasing current-electrode spacings leading to a better depth resolution compared to the Wenner configuration. The number of measurements is larger than for a Wenner survey but smaller than for a Dipole–dipole array. The Wenner–Schlumberger configuration is useful for horizontal and vertical geomorphological structures and can be the best choice as a compromise between the Wenner and Dipole–dipole arrays. The Dipole–dipole array comprises two dipoles formed by the current electrodes on one side and the potential electrodes on the other side. The current and potential spacings are the same and the spacing between them is an integer multiple (n) of the distance (a) between the current and the potential electrodes (Figure 1.2). This array type has a better horizontal resolution, but shallower investigation depth than the Wenner array. Furthermore, its signal-to-noise ratio is smallest and the required number of readings to complete a survey is largest of all three presented configurations.

A comprehensive evaluation of the characteristics of the specific arrays is given in a resistivity tomography tutorial by Loke (2004), together with useful information for the conduction of resistivity surveys and data inversion.

1.3 Data acquisition

Compared to vertical electrical soundings, which typically involve 10 to 20 readings, ERT imaging surveys consist of 100 to several hundred measurements. Depending on the quality of the recorded voltages, the measurements for each electrode configuration have to be repeated until the variance is less than a pre-defined threshold. Acquiring a full ERT data set with about 40 electrodes requires between 0.5 and 1.5 hours (depending on the configuration). The time for data acquisition depends on the number of measured electrode combinations (often called quadripoles) and on the number of repetitions of a single combination.

For the acquisition of the apparent resistivity data sets, multi-electrode systems are commonly used. These systems automatically measure the apparent resistivities for a series of electrode combinations for a given array geometry. Using 40 equally spaced electrodes with a spacing of 5 m and a Wenner array results in a data set of 190 apparent resistivities, a survey line of 195 m length and a penetration depth of about 30 m.

Choice of an appropriate electrode configuration is dependent on the difficult surface conditions associated with mountain regions. Since the maximum current injected into the ground can be quite low, the geometrical factors of the electrode configurations may be critical (Telford *et al.* 1990). For this reason, often Wenner or Wenner–Schlumberger configurations are employed, even though Dipole–dipole configurations may provide superior lateral resolution (Loke 2004). This characteristic is described in more detail in Section 1.5.3.

Further details on different array geometries are given, for instance, in Telford *et al.* (1990) and Reynolds (1997). Applications of different arrays to various geomorphological studies are described in Kneisel (2003, 2006).

1.4 Data processing

Data processing consists basically of applying an appropriate inversion algorithm to the observed apparent resistivity data set to determine the specific resistivity values on a two-dimensional x–z model grid. During the past few years, two- and three-dimensional inversion algorithms for resistivity data have been developed and applied successfully in many environmental and archaeological applications (e.g. Johansson and Dahlin 1996, Mauriello *et al.* 1998, Ogilvy *et al.* 1999, Olayinka and Yaramanci 1999, Daily *et al.* 2004, Günther *et al.* 2006).

The large number of ERT applications in recent years was partly due to the comparatively new availability of 2D and 3D resistivity inversion software like RES2DINV/RES3DINV, which performs a smoothness-constrained inversion using finite difference forward modelling and quasi-Newton inversion techniques (Loke and Barker 1995, 1996). The inversion results in a 2D or 3D specific resistivity model section as opposed to the so-called pseudosections obtained by analysing the apparent resistivities alone. In addition, topography may be incorporated in the inversion, which is an important factor in mountainous glacial and periglacial terrain. Even though there is more than one resistivity inversion software package commercially available, we will explain the exemplary inversion procedure using RES2DINV.

Prior to data processing with the inversion software, it is recommendable to check the data set for abnormally high or low resistivity values. If these values can be attributed to measurement errors and/or bad electrode contact (see Section 1.5.1), they should be excluded manually. The observed apparent resistivity data sets can then be inverted using either the least-squares or the robust inversion scheme (i.e. use of ℓ_2- or ℓ_1-norm for data and model space, respectively; Loke *et al.* 2003). Robust inversion is usually chosen over smooth inversion wherever sharp layer boundaries are expected, as they are reproduced better than with the more smearing least-squares norm.

By default, a homogeneous earth model is used as the starting model, which is obtained by calculating the average of the logarithm of the measured apparent resistivity values. From this resistivity model is calculated a set of apparent resistivities that would be observed in the field if the resistivity model represented the real resistivity distribution in the subsurface. In an iterative algorithm the optimisation method then tries to reduce the difference between the calculated and measured apparent resistivity values by adjusting the resistivity of the model blocks in the resistivity model. A measure of this difference is given by the root-mean-square (RMS) error. By using different starting models the reliability of the inversion results can be tested (Marescot *et al.* 2003).

The least-squares equation is given as

$$p = (J^{\mathrm{T}}J + \lambda C^{\mathrm{T}}C)^{-1}J^{\mathrm{T}}g \qquad (1.2)$$

where p is the model perturbation vector, J is the matrix that includes the sensitivities of the data points with respect to a particular model parameter, g is the discrepancy vector, which contains the differences between measured and calculated apparent resistivities, and ^{T} denotes the transpose of a matrix (Loke and Barker 1995). The matrix C acts as a flatness filter to minimise the under-determined components of the inversion problem and force the inverted models to be smooth. The parameter λ specifies the weighting between data constraints and a-priori information (i.e. the assumed smoothness of the subsurface). Equation (1.2) is solved iteratively (by repeatedly updating g and J) until the RMS of the discrepancies g does not alter significantly after an inversion step and/or it becomes smaller than the measurement accuracy.

In RES2DINV the user can specify the parameter λ, which is called the damping parameter. The higher the damping parameter the smoother the resulting resistivity model, but the weaker the model is constrained by the data set and the larger the RMS error. The lower the damping parameter the noisier the model, but the stronger the data constrain, corresponding to a small RMS error. However, the best model from a geomorphological or geological perspective might not be the one with the lowest possible RMS (see Section 1.5.3). Thus, it is essential to consider the local geomorphological setting in performing the interpretation. This enables unrealistic images of the subsurface structure to be excluded. In order to analyse the results in terms of, for instance, permafrost distribution and characterisation, the final resistivity model has to be interpreted and its reliability assessed. In the following, examples are shown with typical problems associated with resistivity inversion and interpretation.

1.5 Periglacial applications and particularities

1.5.1 Data acquisition

Application of geoelectrical surveys in periglacial environments often implies one major problem, which is the coupling between the electrodes and the sometimes heterogeneous and rocky ground surface. This problem can often be resolved by adding water in the immediate vicinity of the electrodes, by attaching sponges soaked in salt water and/or installing extra electrodes in parallel to the electrodes (see also Chapter 6). Experience has shown that a sufficient supply of water is more important than extra addition of salt. Electrodes should be long (0.4–1 m) and should be firmly positioned between blocks with maximum contact to the ground. Where larger rocks or rock faces are present, small holes can be drilled into the rock using metallic pins or screws as electrodes (Sass 2003, Krautblatter and Hauck 2007). Electrically conductive fluid can be used to further enhance the electrode contact.

The obtained contact resistances depend strongly on the surface conditions and can be as high as several hundred $k\Omega\,m$. Anomalous bad electrode contacts, which may significantly influence the inversion results, are characterised by alternating high and low values in the apparent resistivity pseudosection (so-called W-shaped anomalies). If a faulty electrode is the cause for the anomalies, they are usually visible at all depths and can be readily removed from the data set (Ritz *et al.* 1999). Difficult electrode contact with the ground in ERT surveys can be avoided by using capacitively coupled resistivity systems, which do not require direct (galvanic) contact of electrodes with the subsurface (Hauck and Kneisel 2006, de Pascale *et al.* 2008). Similarly, electromagnetic induction methods are being used to measure the electrical conductivity (1/resistivity) without the necessity of direct contact with the ground (see Chapter 2).

Another source of uncertainty is connected with electrode mislocations along a survey line (Oldenborger *et al.* 2005). It is characterised by the sensitivity of electrical potential to both source and receiver positions. This sensitivity depends not only on source–receiver separation, but also on the location and magnitude of contrasts in electrical resistivity. The resulting errors in the inverted electrical resistivity model due to electrode mislocations can be significant in magnitude with complex spatial distributions and should be avoided. Zhou and Dahlin (2003) showed that the magnitudes of the spacing errors are quite different for the different electrode arrays, being largest for Dipole–dipole, for which a 10% in-line spacing error (i.e. mislocation of an electrode in the direction of the survey line) may cause twice as large an error (>20%) in the observed resistance or apparent resistivity, whereas Wenner and Wenner–Schlumberger arrays give smaller errors (around the magnitude of the position error). Off-line spacing errors (electrode mislocations perpendicular to the survey line) give much smaller errors in the observed apparent resistivity values (one order of magnitude less) (Zhou and Dahlin 2003).

On difficult surface conditions, where well-coupled electrode positions cannot be found at the required location, it is therefore recommended to look for alternative electrode positions perpendicular to the survey line rather than along the survey line. If in-line mislocations cannot be avoided, the Wenner or Wenner–Schlumberger array should be chosen for the measurements.

1.5.2 High resistivities and large resistivity contrasts

Resistivity surveys in mountain areas are often associated with extremely high resistivity values compared to standard ERT surveys. These high resistivity values are caused by a comparatively high ice content or a specific (glacial) origin of the ice, but may be influenced to a substantial degree by the choice of the inversion parameter, the measurement geometry or ambiguities due to the principle of equivalence (e.g. Telford *et al.* 1990, Hauck and Vonder Mühll 2003b). Ground ice detection on rock glaciers or ice-cored moraines has been one of the major applications of ERT in mountain permafrost studies in recent years (Vonder Mühll *et al.* 2000, Ishikawa *et al.* 2001, Hauck *et al.* 2003, Marescot *et al.* 2003, Kneisel 2004, 2006, Farbrot *et al.* 2005, Maurer and Hauck 2007). Three questions are usually discussed: (i) the occurrence and extent of possible ground ice, (ii) the ice content and (iii) the origin of the ice (e.g. Haeberli and Vonder Mühll 1996, Ishikawa *et al.* 2001, Kneisel 2004). In the following, field results from two active rock glaciers in the Eastern Alps are compared: rock glaciers Murtel (Upper Engadine, Swiss Alps) and Stelvio (Italian Alps). The second question is also specifically addressed in Chapters 9 and 10, the third question is addressed in Chapter 6.

Figure 1.3a shows the inversion results for a longitudinal ERT profile across the tongue of rock glacier Murtel. Electrode coupling was especially difficult as the surface consists of big blocks (up to 2 m high) without fine material. The survey line leads from near a 52 m borehole drilled in 1987 (horizontal distance − 80) downslope across the tongue of the rock glacier to the non-permafrost regions below. The maximum resistivities of 2 MΩ m are located in the depth range between 5 and 15 m, corresponding to the massive ice found in the drill core (Haeberli and Vonder Mühll 1996). Underneath, the resistivity decreases again, representing the decrease in ice content as the amount of sand and coarse-grained debris becomes larger. The sharp gradient at the tongue (shown by the dashed line) is clearly visible, delineating the extent of the ice body. Resistivities in the permafrost-free area in front of the rock glacier are less than 5 kΩ m.

Figure 1.3b shows the inversion results for a similar survey along a rock glacier near Stelvio Pass, Italy (Hauck and Vonder Mühll 2003b). The survey was conducted along a 200 m line reaching from the main part of the rock glacier (horizontal distance −100) to the tongue (horizontal distance 100). The resistivity maximum (region A in Figure 1.3b) of the inversion model, corresponding to the ice body, is thinner and has lower values (up to 500 kΩ m) than in the case of rock glacier Murtel. In the frontal part (region B) and at greater depth (region C), resistivity values are less than 50 kΩ m with sharp resistivity gradients at 10–25 m depth and near horizontal distance 30 (indicated by the black line).

In both cases high resistivities ($>10^5$ Ω m) indicate the presence of frozen material and sharply decreasing resistivities indicate the horizontal and vertical extents of the presumed ice bodies. In contrast to rock glacier Murtel, where the resistivities can be related to the drill core, no ground truth is available at rock glacier Stelvio. Geomorphological and vegetational analyses (Cannone *et al.* 2003) indicate an active part of the rock glacier corresponding to region A and a less active or inactive part corresponding to region B. These findings are supported by the resistivity contrast of one order of magnitude between regions A and B. The large resistivity decrease with depth between regions A and C and the comparatively low resistivity values in region C are interpreted as the lower boundary of the ice-rich layer.

However, using the depth-of-investigation (DOI) method, where inversion results from different starting models are compared, Marescot *et al.* (2003) showed that in the case of large resistivity gradients the reliability of the inversion

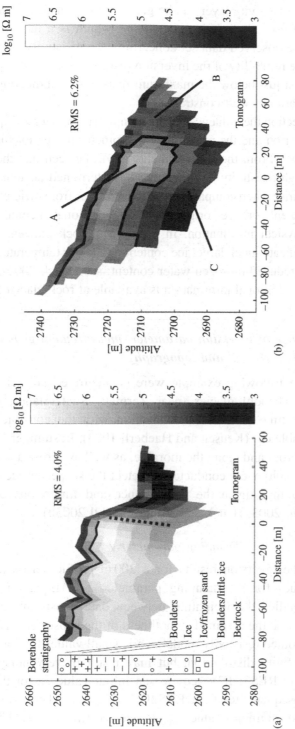

Figure 1.3.   Model inversion results for two rock glaciers in the Eastern Alps: (a) Murtel, Upper Engadin, (b) Stelvio, Italian Alps (from Hauck and Vonder Mühll 2003b).

results below a high-resistivity layer can be greatly diminished. In a similar way, Hauck *et al.* (2003) used the relative sensitivity, which is a measure of information content in the observed data set concerning the resistivity of each model block, to estimate the reliability of the inversion results. For both rock glaciers the sensitivity was lowest just below the maximum resistivities, indicating that these vertical boundaries are not well constrained.

In the following section the influences of the damping parameter λ (Equation 1.2), the convergence criterion and the measurement geometry on the maximum resistivity values are shown. From these experiments it will be seen that the resistivity values within the high-resistivity zones cannot be determined accurately. Consequently, difficulties arise when comparing resistivity values from different field sites concerning ice content or ice origin without additional geomorphological, geological or geophysical information. In the case of rock glaciers Murtel and Stelvio the results indicate much larger ice content or a lower temperature at Murtel (corresponding to a reduced unfrozen water content: see Hauck 2002), but uncertainty remains as no additional information is available at rock glacier Stelvio.

1.5.3 Influence of inversion parameter, measurement geometry and topography

Field surveys of the following example were concentrated around the eastern lateral moraine of Theodul glacier (near Zermatt, Switzerland), which was studied for the first time in 1974 in connection with the construction of the Klein-Matterhorn cable car (Keusen and Haeberli 1983). In summer 1998, 200 m long ERT profiles across and along the moraine, as well as across a nearby valley with complex topography were conducted to detect the suspected ice core within the moraine and to investigate the performance and limitations of the ERT method (Hauck *et al.* 2003, Hauck and Vonder Mühll 2003b).

Damping parameter λ

Figure 1.4 shows the inversion results for a 200 m profile across the moraine using different values for the damping parameter λ in Equation (1.2). This damping factor specifies the weighting between data constraints and a-priori information, in this case the assumed smoothness of the subsurface. Large values for λ result in a smoothed resistivity model, while small values for λ allow a more data-consistent resistivity distribution, but may result in very noisy resistivity models. The program RES2DINV requires an initial value λ_0 for the first iteration. For each subsequent iteration, RES2DINV reduces the λ value by 50% until a user-specified minimum value λ_{min} is reached (in our case 1/5 of λ_0).

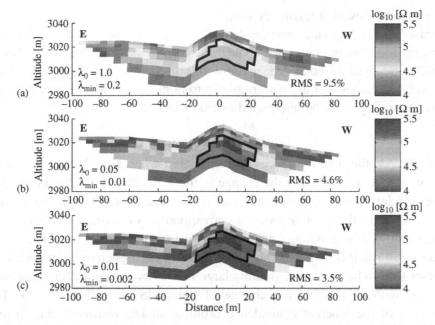

Figure 1.4. Tomographic inversions of data set recorded on ice-cored moraine near Zermatt for (a) large, (b) intermediate and (c) small damping values. The resistive zone interpreted to be the ice core is marked by a solid line (from Hauck and Vonder Mühll 2003b). For colour version see Plate 1.

In Figure 1.4 (Plate 1), the model with a large value for λ is comparatively smooth but has a high RMS error (9.5%) (Figure 1.4a), while the model with a low value for λ contains more noise but has a lower RMS error (3.5%) (Figure 1.4c). An optimal value for λ can be determined using a series of inversion runs with different values for λ, and choosing the result where the RMS error is minimal, but not much smaller than the estimated measurement accuracy (in our case 5%) (Hauck *et al.* 2003). In this way, a trade-off between data consistency and smoothing can be achieved (Figure 1.4b).

Within the uppermost 10 m of the tomograms, the resistivity distribution is fairly irregular, representing the active layer and/or sedimentary material with only low amounts of ice. At greater depths, a pronounced high-resistivity zone is observed in the centre of all tomograms (delimited by the solid black line). The boundaries of this zone are determined on the basis of the tomogram in Figure 1.4b. The maximum resistivity values within this zone vary from $50\,k\Omega\,m$ in Figure 1.4a up to $500\,k\Omega\,m$ in Figure 1.4c. Similarly, the resistivity contrast between the high-resistivity anomaly and the host material is lowest for the highest degree of smoothing (Figure 1.4a), but its general shape is similar in all tomograms. This anomalous zone is therefore interpreted as the ice core of the moraine.

The results show that resistivity values and even resistivity contrasts between anomalies and background material depend strongly on the λ-value used during inversion. Joint interpretations of tomograms computed with different λ-values are often required to obtain meaningful interpretations (Hauck *et al.* 2003). However, anomalies that are discernible in tomograms computed with different parameter settings may be interpreted as true features.

Convergence criterion

In RES2DINV the iteration process is terminated when a predefined maximum iteration number is reached or the change in the RMS error from one iteration to the next is lower than a predefined limit (Loke and Barker 1996). However, a small RMS error does not necessarily correspond to a realistic inversion model result, because a large number of iterations will tend to overfit the data, so that artefacts will result from inversions of the data errors. Especially large resistivity contrasts tend to be increased, yielding large resistivity variations for subsequent iterations without a significant change of the RMS error (Figure 1.5). This expression of the inherent equivalence problem in DC resistivity data further complicates the interpretation of high resistivity values in inversion models, because the longer the iteration process, the higher the maximum resistivities.

Electrode configuration and measurement geometry

In addition to the above inversion parameter, the chosen electrode configuration can also influence the inversion results. Figure 1.6 shows results from measurements with both Wenner and Dipole–dipole arrays at a sporadic permafrost

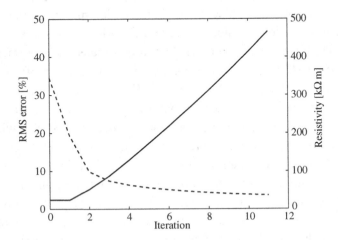

Figure 1.5.   Evolution of the root-mean-square error (RMS) (dashed line) and corresponding maximum model resistivity (solid line) during the iteration process (from Hauck and Vonder Mühll 2003b).

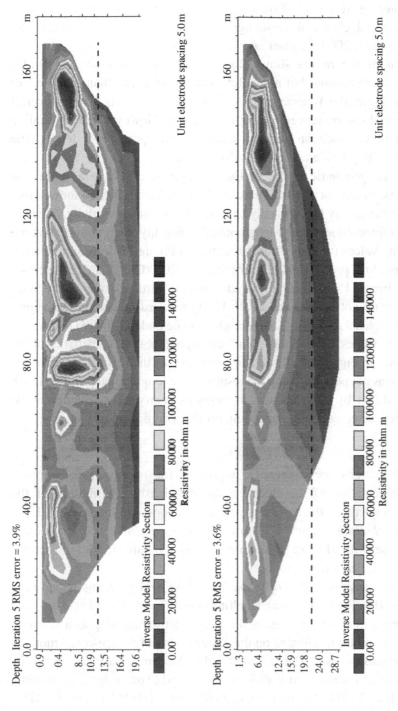

Figure 1.6. Comparison of two resistivity tomograms performed at the same location: Dipole–dipole (upper panel) and Wenner survey (bottom panel) on a scree slope with isolated permafrost lenses. Electrode spacing is 5 m.

17

site in the Bever Valley, Swiss Alps. Areas with resistivity values as high as 120 kΩ m are interpreted as permafrost lenses, which are in good agreement with findings from vertical electrical soundings and refraction seismic tomography surveys (Kneisel *et al.* 2000, Kneisel and Hauck 2003).

A comparison of the results shows that the Dipole–dipole array provides superior horizontal resolution, but a smaller penetration depth than the Wenner array. In this example, the Wenner results seem to be sufficient for the overall detection of permafrost presence/absence at this site. However, for a detailed characterisation of the location and the extent of the permafrost lenses, the Dipole–dipole results provide additional information.

Furthermore, the geometric layout of the survey line can have a significant influence on the measurement results. Hauck and Vonder Mühll (2003b) showed that the orientation of the survey line critically changes the absolute resistivity values in the case of strong three-dimensional topographical variability. They showed that the absolute resistivity values for the ice core presented in Figure 1.4 were significantly higher for a longitudinal profile along the moraine top (200 kΩ m) than for the cross-profile shown in Figure 1.4 (110 kΩ m). In addition, the vertical extent of the high-resistivity zone could not be detected at all. Using forward modelling techniques they showed that isolated resistivity anomalies tend to be underestimated in terms of absolute resistivity values. On the other hand, the topographic characteristics of the cross-profile shown in Figure 1.4 allowed an estimation of the vertical extent of the ice body, which was not possible using the results of the longitudinal profile (Hauck and Vonder Mühll 2003b). Consequently, the array geometry and orientation of the profile have to be chosen carefully depending on the specific aim of the survey.

Topography

Topographic corrections are especially important for regions with complex terrain, as the calculation of the electrical current density distribution of the subsurface is dependent on the surface topography (Tong and Yang 1990, Loke 2000, Günther *et al.* 2006). Figure 1.7 shows the resistivity inversion results of a 200 m long profile with (Figure 1.7a) and without (Figure 1.7b) topographic corrections during inversion.

This survey line was conducted over extreme altitudinal gradients including a downslope region (I in Figure 1.7) with an altitudinal difference of 50 m, a crossing of a small sediment-filled valley (region II), a 20 m altitudinal step to a flat rock plateau (region III) until it terminated on the eastern flank of the ice-cored moraine (region IV) discussed above. The different subsurface materials can easily be distinguished by their respective resistivity values (3–6 kΩ m: bedrock underneath the slope (I and II), 10–20 kΩ m: rock plateau (III) and >100 kΩ m: ice core (IV); Hauck and Vonder Mühll 2003b).

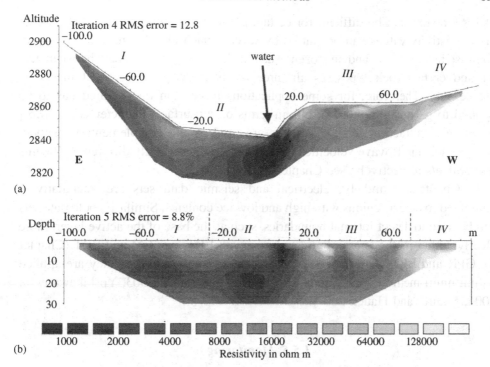

Figure 1.7. Model inversion results for a survey line over complex geomorphological terrain near Zermatt, Switzerland: (a) with and (b) without topographic corrections. Regions I–IV correspond to features explained in the text (from Hauck and Vonder Mühll 2003b).

Comparing Figures 1.7a and 1.7b large differences can be found, especially in region III and near the surface water flow at horizontal distance 5 m. As can be seen from Figure 1.7b, an inversion without incorporated topography would yield erroneous results leading to possible misinterpretations of geological and periglacial features. For example, the low-resistivity zone at 10–20 m depth between horizontal distances −90 m and 60 m in Figure 1.7b could be misinterpreted as a permafrost-free layer, whereas these low resistivity values are solely due to inversion artefacts induced by the rather confined conductive zone of water flow seen in the vicinity of horizontal distance 5 m in Figure 1.7a. A large influence of topography on the inversion results is typical for complex terrain and heterogeneous subsurface characteristics.

1.5.4 Non-uniqueness of the interpretation of high resistivities due to air, ice and bedrock

Even though ERT is well suited to detect ground ice occurrences, due to the strong dependence of electrical resistivity on unfrozen water/ice content, the interpretation of ERT results can be difficult for certain permafrost environments, as the measured resistivity values can be caused by several materials. Whereas the resistivity contrast between ice and unfrozen water is huge, it can be small between ice, air and certain rock types, as all three behave almost as electrical insulators (Table 1.1). Therefore, for some applications more than one method has to be applied to get unambiguous results in terms of subsurface characterisation. Most commonly, refraction seismic surveys are used as the complementary method, because seismic P-wave velocities for ice and air are markedly different (3500 m/s and 330 m/s respectively, see Chapters 3, 9 and 10).

In Chapters 9 and 10, electrical and seismic data sets are quantitatively combined to detect regions with high and low ice contents. Similarly, both data sets can be used to detect internal boundaries, such as the base of the active layer, more clearly. Other methods are often combined with ERT, e.g. EM and ERT (Chapter 8), GPR and ERT (Farbrot et al. 2005, Otto and Sass 2006) or they are applied within multi-method applications (e.g. Kneisel and Hauck 2003, Yoshikawa et al. 2006, Maurer and Hauck 2007).

1.5.5 Monitoring

For the monitoring of time-dependent processes (so-called time-lapse experiments) 2D resistivity surveys are repeated over the same survey line at different times (e.g. Barker and Moore 1998, French et al. 2002, Hauck 2002, Müller et al. 2003, French and Binley 2004, Kneisel 2006, Rings et al. 2008, Hilbich et al. 2008). This requires accurate and reproducible results even on rough terrain. Fixed-electrode arrays can be used to minimise changes due to temporally varying electrode locations and contacts. Hauck (2002) introduced a semi-automated survey design, on a mountain slope with fine-grained weathered material overlying bedrock, where a manual switchbox is attached to the electrode array via permanently installed cables to enable measurements throughout the year (even in the presence of thick snow cover). The currently seven-year-long ERT monitoring time series from the Schilthorn, Swiss Alps, shows the applicability of the approach on daily, seasonal and interannual time scales (Hilbich et al. 2008).

As an example, Figure 1.8 shows results from test measurements on a blocky moraine in the Muragl glacier forefield using a fixed electrode array with 2 m electrode spacing. In consideration of the rugged surface conditions, the inversion results are of good quality and reproducible. The data were first inverted independently. This approach has given only limited information on subsurface resistivity changes with time since the inversion routine tries to minimise the difference between the measured and calculated apparent resistivity values. Thus,

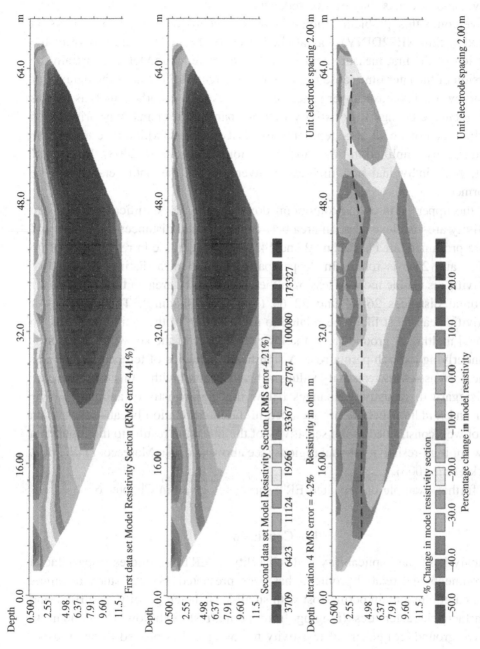

Figure 1.8. Resistivity tomograms of a Wenner–Schlumberger survey on a moraine consisting partly of coarse debris. Upper panel represents the survey on 4 August, middle panel the survey on 27 August, bottom panel the percentage change in resistivity. Dashed line indicates inferred depth of active layer.

21

differences in the resulting resistivity values of the two individual inversions are not necessarily related to actual changes in the subsurface resistivity distribution, as inversion artefacts may evolve independently in the two separate inversions. To overcome this problem a joint inversion technique (time-lapse inversion method within RES2DINV) was applied. Hence, the model obtained from the inversion of the first measurement is used as a reference model to constrain the inversion of the later time-lapse measurements. Figure 1.8 displays the data set of the surveys on 4 August (upper panel) and on 27 August (middle panel) as well as the percentage change in resistivity (bottom panel). Problems may arise when inverting data sets with large temporal resistivity changes, which tend to become unrealistically amplified in time-lapse inversions (Rings *et al.* 2008). In cases of doubt, both individual and time-lapse inversions of the data sets should be performed.

In the upper parts of the tomogram down to 3 m, no significant changes in resistivity are visible, except an area between horizontal distances 28 and 30 and a more prominent area between 52 and 64, where a decrease in resistivities of up to 20% and 25% is found. In deeper parts below 3 m, a distinct decrease of resistivities is visible more or less over the whole survey area and especially over horizontal distances 26–44 and 52–64 (light grey shading). The decrease of resistivities can most likely be related to a thickening of the active layer due to seasonal melting of ground ice at the lower boundary of the active layer towards the underlying ice-rich permafrost. A significant influence of temperature effects on the decrease of the resistivity values or heavy rainfall within this time span can be disregarded. Resistivity changes below the high-resistivity anomaly in the upper and middle panel are not included in the interpretation because these data are not well constrained as the sensitivity of the inversion results to the input data is low for high-resistivity model regions (see above and also Marescot *et al.* 2003, Hauck *et al.* 2003).

A further example on repeated ERT surveys is given in Chapter 6.

1.6 Conclusions

An analysis of the applicability and reliability of ERT for studies in periglacial environments and related problems has been presented. For the study of alpine and subarctic mountain permafrost with small-scale heterogeneity of surface and subsurface characteristics (ranging from permafrost with low ice content to massive ground ice) electrical resistivity tomography has proved to be an especially well suited and multi-functional method. ERT is a comparatively fast method to image the subsurface and infer permafrost characteristics even on rugged alpine terrain with rough surface conditions (see e.g. Kneisel and Kääb

2007). Ambiguities in resistivity model inversion and interpretation include the model dependence on inversion parameters such as the number of iterations and the damping factor, influence of topography and measurement geometry, as well as misinterpretations of high resistivity values caused by air-filled cavities. The reliability of the models and possible interpretations were discussed for a variety of applications in mountain permafrost research. Key results from these studies include:

- In the case of large resistivity contrasts due to comparatively high ice contents, a reliable distinction between permafrost and non-permafrost areas is easily possible. The spatial extent of ice in rock glaciers, ice-cored moraines and other ice-rich permafrost occurrences can be determined. However, the reliability of inversion results at greater depths has to be assessed (relative sensitivity; depth-of-investigation method).
- Choice of appropriate inversion parameter can be important especially for noisy data sets. Multiple inversions using, for instance, different values for the damping parameter (model smoothing) may help to distinguish model artefacts from real anomalies.
- The convergence criterion should be set according to the estimated error level in the observed data. An unnecessarily prolonged iteration process increases the modelled resistivity contrasts and the resulting maximum resistivities.
- Model studies with synthetic data sets show an underestimation of resistivity maxima for isolated anomalies and the Wenner array. Additionally, the depth of a high-resistivity anomaly is generally predicted as too shallow in the model results.
- Different array geometries of the same profile may lead to significant differences in the model results. The Wenner array has larger penetration depth and leads to less noisy models, but has comparatively low lateral resolution. The Dipole–dipole array may provide improved resolution, but leads to noisier models. In regions with strong three-dimensional variability, at least two surveys at right angles should be conducted.
- Electrical resistivity tomography surveys can be conducted on very steep and complex terrain. However, the topography must be incorporated prior to inversion.
- Complementary field measurements using refraction seismic tomography showed that care has to be taken when interpreting high-resistivity anomalies without additional data. Seismic P-wave velocities can be used to distinguish between resistivity anomalies caused by air, ice or other materials (see also Chapters 9 and 10).
- Repeated ERT measurements along the same survey are used to monitor resistivity changes with time, which can be connected to freezing and thawing processes in the subsurface. Additional causes for resistivity changes are water input through rain or snowmelt, and desaturation due to evaporation or water outflow (see also Chapter 6).
- Up to now an efficient 3D geophysical mapping of the subsurface in mountainous environments with rough terrain on a larger scale has not been possible. As a realistic compromise, results of several 2D geophysical surveys at close distance can be merged to build up a pseudo 3D image of the subsurface characteristics and lithology, an approach that is currently in progress in several studies in the Swiss Alps and that will become increasingly important in the near future.

1.7 Checklist

- Fieldwork should be thoroughly planned especially in remote high mountain areas, e.g. estimation of time available for the field measurements (efficiently using good weather conditions) and battery capacity.
- What is the minimum expected number of profiles/surveys?
- Is the equipment in working order, are the batteries charged?
- Always bring spare batteries if possible.
- Choose layout and electrode configuration according to estimated complexity of subsurface layering and ground resistance – compose and/or upload configuration files for the planned measurements.
- Perform contact resistance check for each electrode.
- In the case of rough surface conditions with high ground resistance, use water or sponges soaked with water attached to the electrodes to improve electrical contact – another possibility is to use additional electrodes in parallel.
- Is there any information on the expected subsurface conditions based on geomorphological and geological setting and interpretation?
- Is there any information about expected depth of layers?
- Is there any possibility of shallow layers that might not be resolved?
- Are there any other critical factors (air-filled voids, water flow at the electrodes etc.) influencing the current flow?
- Note electrodes that are badly coupled.
- Note topography for the following interpretation of the data.
- Make whenever possible or necessary for the scientific question cross-checks using complementary geophysical methods.
- Note location of the electrodes for possible repetition or time-lapse measurements (monitoring).
- After the survey: recharge batteries and save data.

REFERENCES

Barker, R. D. (1989). Depth of investigation of collinear symmetrical four electrode arrays. *Geophysics*, **54**(8), 1031–1037.

Barker, R. D. and Moore, J. (1998). The application of time-lapse electrical tomography in groundwater studies. *The Leading Edge*, **17**, 1454–1458.

Cannone, N., Guglielmin, M., Hauck, C. and Vonder Mühll, D. (2003). The impact of recent glacier fluctuation and human activities on the permafrost distribution: a case study from Stelvio Pass (Italian Central-Eastern Alps). *Proceedings of the 8th International Conference on Permafrost*, Zürich, Switzerland, 125–130.

Daily, W., Ramirez, A., Binley, A. and LaBrecque, D. (2004). Electrical resistance tomography. *The Leading Edge*, **23**, 438–442.

de Pascale, G. P., Pollard, W. H. and Williams, K. K. (2008). Geophysical mapping of ground ice using a combination of capacitive coupled resistivity and ground-penetrating radar, NWT, Canada. *Journal of Geophysical Research*, **113**, F02590.

Farbrot, H., Isaksen, K., Eiken, T., Kääb, A. and Sollid, J. L. (2005). Composition and internal structures of a rock glacier on the strandflat of western Spitsbergen, Svalbard. *Norsk Geografisk Tidsskrift*, **59**(2), 139–148.

French, H. K. and Binley, A. (2004). Snowmelt infiltration: monitoring temporal and spatial variability using time-lapse electrical resistivity. *Journal of Hydrology*, **297**, 174–186.

French, H. K., Hardbattle, C., Binley, A., Winship, P. and Jakobsen, L. (2002). Monitoring snowmelt induced unsaturated flow and transport using electrical resistivity tomography. *Journal of Hydrology*, **267**(3–4), 273–284.

Günther, T., Rücker, C. and Spitzer, K. (2006). Three-dimensional modelling and inversion of dc resistivity data incorporating topography – II. Inversion. *Geophysical Journal International*, **166**, 506–517.

Haeberli, W. and Vonder Mühll, D. (1996). On the characteristics and possible origins of ice in rock glacier permafrost. *Zeitschrift für Geomorphologie, Supplement*, **104**, 43–57.

Hauck, C. (2002). Frozen ground monitoring using DC resistivity tomography. *Geophysical Research Letters*, **29**(21), 2016.

Hauck, C. and Kneisel, C. (2006). Application of capacitively-coupled and DC electrical resistivity imaging for mountain permafrost studies. *Permafrost and Periglacial Processes*, **17**(2), 169–177.

Hauck, C. and Vonder Mühll, D. (2003a). Evaluation of geophysical techniques for application in mountain permafrost studies. *Zeitschrift für Geomorphologie, Supplement*, **132**, 161–190.

Hauck, C. and Vonder Mühll, D. (2003b). Inversion and interpretation of 2-dimensional geoelectrical measurements for detecting permafrost in mountainous regions. *Permafrost and Periglacial Processes*, **14**(4), 305–318.

Hauck C., Vonder Mühll, D. and Maurer, H. (2003). Using DC resistivity tomography to detect and characterise mountain permafrost. *Geophysical Prospecting*, **51**, 273–284.

Hilbich, C., Hauck, C., Scherler, M., Schudel, L., Völksch, I., Hoelzle, M., Vonder Mühll, D. and Mäusbacher, R. (2008). Monitoring mountain permafrost evolution using electrical resistivity tomography: A 7-year study of seasonal, annual, and long-term variations at Schilthorn, Swiss Alps. *Journal of Geophysical Research* **113**, F01590, doi:10.1029/2007JF00799.

Hoekstra, P. and McNeill, D. (1973). Electromagnetic probing of permafrost. *Proceedings of the 2nd International Conference on Permafrost*, Yakutsk, Siberia, 517–526.

Ishikawa, M., Watanabe, T. and Nakamura, N. (2001). Genetic difference of rock glaciers and the discontinuous mountain permafrost zone in Kanchanjunga Himal, Eastern Nepal. *Permafrost and Periglacial Processes*, **12**(3), 243–253.

Johansson, S. and Dahlin, T. (1996). Seepage monitoring in an earth embankment dam by repeated resistivity measurements. *European Journal of Environmental and Engineering Geophys*ics, **1**, 229–247.

Keusen, H. R. and Haeberli, W. (1983). Site investigation and foundation design aspects of cable car construction in Alpine permafrost at the "Chli Matterhorn", Wallis, Swiss Alps. *Proceedings of the 4th International Conference on Permafrost*, Fairbanks, Alaska, 601–605.

Kneisel, C. (1999). *Permafrost in Gletschervorfeldern – Eine vergleichende Untersuchung in den Ostschweizer Alpen und Nordschweden*. Trierer Geographische Studien, 22, 156pp.

Kneisel, C. (2003). Electrical resistivity tomography as a tool for geomorphological investigations – some case studies. *Zeitschrift für Geomorphologie, Supplement*, **132**, 37–49.

Kneisel, C. (2004). New insights into mountain permafrost occurrence and characteristics in glacier forefields at high altitude through the application of 2D resistivity imaging. *Permafrost and Periglacial Processes*, **15**, 221–227.

Kneisel, C. (2006). Assessment of subsurface lithology in mountain environments using 2D resistivity imaging. *Geomorphology*, **80**, 32–44.

Kneisel, C. and Hauck, C. (2003). Multi-method geophysical investigation of an isolated permafrost occurrence. *Zeitschrift für Geomorphologie, Supplement*, **132**, 145–159.

Kneisel, C. and Kääb, A. (2007). Mountain permafrost dynamics within a recently exposed glacier forefield inferred by a combined geomorphological, geophysical and photogrammetrical approach. *Earth Surface Processes and Landforms* **32**(12), 1797–1810.

Kneisel, C., Hauck, C. and Vonder Mühll, D. (2000). Permafrost below the timberline confirmed and characterized by geoelectrical resistivity measurements, Bever Valley, eastern Swiss Alps. *Permafrost and Periglacial Processes*, **11**, 295–304.

Kneisel, C., Beylich, A. A. and Sæmundsson, T. (2007). Reconnaissance surveys of contemporary permafrost environments in central Iceland using geoelectrical methods: implications for permafrost degradation and sediment fluxes. *Geografiska Annaler*, **89**, 41–50.

Krautblatter, M. and Hauck, C. (2007). Electrical resistivity tomography monitoring of permafrost in solid rock walls. *Journal of Geophysical Research*, **112**, F02S20, doi:10.1029/2006JF000546.

Kulessa, B., Hubbard, B. and Brown, G. H. (2003). Cross-coupled flow modeling of coincident streaming and electrochemical potentials and application to subglacial self-potential data. *Journal of Geophysical Research*, **108**(B8), 2381, doi:10.1029/2001JB001167.

Loke, M. H. (2000). Topographic modelling in electrical imaging inversion. *Proceedings of the 62nd EAGE Conference & Technical Exhibition*, Glasgow, **D02**, 4pp.

Loke, M. H. (2004). Lecture notes on 2D and 3D electrical imaging surveys. Available in pdf-format on www.geoelectrical.com.

Loke, M. H. and Barker, R. D. (1995). Least-squares deconvolution of apparent resistivity. *Geophysics*, **60**, 1682–1690.

Loke, M. H. and Barker, R. D. (1996). Rapid least-squares inversion of apparent resistivity pseudosections using a quasi-Newton method. *Geophysical Prospecting*, **44**, 131–152.

Loke, M. H., Acworth, I. and Dahlin, T. (2003). A comparison of smooth and blocky inversion methods in 2D electrical imaging surveys. *Exploration Geophysics*, **34**, 182–187.

Marescot, L., Loke, M. H., Chapellier, D., Delaloye, R., Lambiel, C. and Reynard, E. (2003). Assessing reliability of 2D resistivity imaging in permafrost and rock glacier studies using the depth of investigation index method. *Near Surface Geophysics*, **1**(2), 57–67.

Maurer, H. and Hauck, C. (2007). Geophysical imaging of alpine rock glaciers. *Journal of Glaciology*, **53**(180), 110–120.

Mauriello, P., Monna, D. and Patella, D. (1998). 3D geoelectric tomography and archaeological applications. *Geophysical Prospecting*, **46**(5), 543–570.

Müller, M., Mohnke, O., Schmalholz, J. and Yaramanci, U. (2003). Moisture assessment with small-scale geophysics – The Interurban Project. *Near Surface Geophysics*, **1**(4), 173–182.

Ogilvy, R., Meldrum, P. and Chambers, J. (1999). Imaging of industrial waste deposits and buried quarry geometry by 3-D resistivity tomography. *European Journal of Environmental and Engineering Geophysics*, **3**, 103–113.

Olayinka, A. and Yaramanci, U. (1999). Choice of the best model in 2-D geoelectrical imaging: Case study from a waste dump site. *European Journal of Environmental and Engineering Geophysics*, **3**, 221–244.

Oldenborger, G. A., Routh, P. S. and Knoll, M. D. (2005). Sensitivity of electrical resistivity tomography data to electrode position errors. *Geophysical Journal International*, **163**(1), 1–9.

Otto, J. C. and Sass, O. (2006). Comparing geophysical methods for talus slope investigations in the Turtmann valley (Swiss Alps). *Geomorphology*, **76**(3–4), 257–272.

Reynolds, J. M. (1997). *An Introduction to Applied and Environmental Geophysics*. John Wiley & Sons.

Rings, J., Scheuermann, A., Preko, K. and Hauck, C. (2008). Soil water content monitoring on a dike model using electrical resistivity tomography. *Near Surface Geophysics*, **6**, 123–132.

Ritz, M., Robain, H., Pervago, E., Albouy, Y., Camerlynck, C., Descloitres, M. and Mariko, A. (1999). Improvement to resistivity pseudosection modelling by removal of near-surface inhomogeneity effects: application to a soil system in south Cameroon. *Geophysical Prospecting*, **47**, 85–101.

Sass, O. (2003). Moisture distribution in rockwalls derived from 2D-resistivity measurements, *Zeitschrift für Geomorphologie, Supplement*, **132**, 51–69.

Scott, W., Sellmann, P. and Hunter, J. (1990). Geophysics in the study of permafrost. In *Geotechnical and Environmental Geophysics*, ed. Ward, S., Society of Exploration Geophysics, Tulsa, pp. 355–384.

Slater, L. and Lesmes, D. (2002). IP interpretation in environmental investigations. *Geophysics*, **67**(1), 77–88.

Telford, W. M., Geldart, L. P. and Sheriff, R. E. (1990). *Applied Geophysics*. Cambridge University Press.

Tong, L. and Yang, C. (1990). Incorporation of topography into two-dimensional resistivity inversion. *Geophysics*, **55**, 354–361.

Vonder Mühll, D., Hauck, C. and Lehmann, F. (2000). Verification of geophysical models in Alpine permafrost using borehole information. *Annals of Glaciology*, **31**, 300–306.

Weller, A. and Börner, F. (1996). Measurements of spectral induced polarization for environmental purposes. *Environmental Geology*, **27**, 329–334.

Yoshikawa, K., Leuschen, C., Ikeda, A., Harada, K., Gogineni, P., Hoekstra, P., Hinzman, L., Sawada, Y. and Matsuoka, N. (2006). Comparison of geophysical investigations for detection of massive ground ice (pingo ice). *Journal of Geophysical Research*, **111**(E6), CiteID E06S19.

Zhou, B. and Dahlin, T. (2003). Properties and effects of measurement errors on 2D resistivity imaging surveying. *Near Surface Geophysics*, **1**, 105–117.

2

Electromagnetic methods

A. Hördt and C. Hauck

2.1 Introduction

Electromagnetic (EM) methods have been widely used in studies of lowland arctic permafrost (e.g. Hoekstra and McNeill 1973, Sartorelli and French 1982, Rozenberg *et al.* 1985, Harada *et al.* 2000) and studies of sea ice thickness determination (Haas 2004), including airborne applications (Fraser 1978, Pfaffling *et al.* 2004). Applications of EM methods in mountainous regions are less frequent (Schmöller and Frühwirth 1996, Hauck *et al.* 2001, Beylich *et al.* 2003, Bucki *et al.* 2004, Maurer and Hauck 2007), but have been increasing in recent years. Electromagnetic techniques include frequency-domain EM systems (FEM), time-domain electromagnetic systems (TDEM), systems using very low frequencies (VLF) and the so-called radiomagnetotelluric method (RMT).

Similar to the electrical methods (see Chapter 1) the physical parameter allowing a differentiation between ice and water or frozen and unfrozen substratum is the electrical resistivity (in ohm metres, Ωm) or more commonly its reciprocal, the electrical conductivity (in siemens/metre or usually millisiemens/metre, mS/m). A marked increase in resistivity (or decrease in conductivity) with decreasing temperature near the freezing point has been shown in many previous field studies (e.g. Hoekstra *et al.* 1975, Seguin 1978, Rozenberg *et al.* 1985) and laboratory studies (e.g. Olhoeft 1978, Pandit and King 1978, King *et al.* 1988, see also Chapter 7). For the major part of the commonly used frequency range, this characteristic physical property is independent of the measurement electromagnetic frequency; however, the absolute conductivity values are frequency dependent and differ strongly for different materials and unfrozen pore water contents (Hoekstra and McNeill 1973, Olhoeft 1978).

Applied Geophysics in Periglacial Environments, eds. C. Hauck and C. Kneisel. Published by Cambridge University Press. © Cambridge University Press 2008.

In the following, the different measurement principles are introduced and discussed concerning their suitability for periglacial applications.

2.2 Background

2.2.1 Measurement principle

Electromagnetic methods are based on electric currents that are induced in the earth by a time-varying current in a transmitter (Figure 2.1). The currents in the earth depend on the electrical conductivity distribution in the subsurface, and cause their own EM field called the secondary field, which superimposes on the primary field generated by the respective instrument transmitter. From the EM field measured by a receiver, the spatial distribution of electrical conductivity may be determined. As indicated in Figure 2.1, loops or coils are commonly used as transmitters or receivers. The confined body used in the sketch may be replaced by a conducting halfspace (a term used to describe the non-conducting air layer over a conducting earth), or any other conductivity distribution, in which case the current flow will be more complicated.

Depending on the time variation of the transmitter current, EM methods operate either in the frequency domain or in the time domain. Frequency-domain electromagnetic (FEM) methods use a sinusoidal current with a specific single frequency at a time. The signal observed at the receiving sensor, which has the

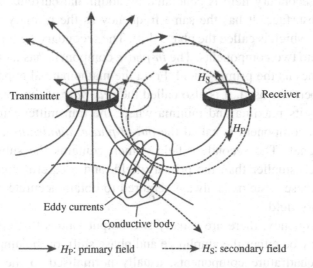

Figure 2.1. Principle of electromagnetic methods (after Militzer and Weber 1985). The varying magnetic field of the transmitter induces currents in the conductor. The induced currents have a secondary field that is superimposed on the primary field.

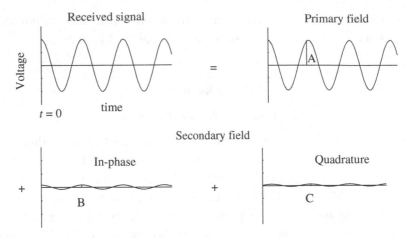

Figure 2.2. Schematic sketch of the different components of the output signal. The signal consists of a primary field, which is usually much larger than the secondary field. The secondary field can be decomposed into a part that is *in-phase* with the primary field, and the out-of-phase component, called *quadrature*. The output is the amplitude of secondary in-phase and quadrature amplitudes, normalised with the primary field.

same frequency as the transmitter current, can be decomposed into primary and secondary fields (Figure 2.2). The primary field is generated only by the transmitter current, and thus it would be measured in the absence of any conducting material. The secondary field is generated by additional currents induced in the conducting subsurface. It has the same frequency as the primary field, but lags behind in time, which is called the phase shift. The secondary field may be further decomposed into two components. The *in-phase* component has its zero crossings at the same times as the primary field. From the mathematical expression, where complex numbers are used, it is also called the *real part* of the signal. The other component has its maxima and minima where the transmitter current has zero crossings. This component is called the *out-of-phase*, *quadrature*, or *imaginary part* of the signal. The secondary field, which contains the subsurface information, is much smaller than the primary field, and a careful compensation or calibration of these systems is always required to obtain accurate measurements of the secondary field.

For each frequency, there are usually two output values that characterise the signal. This may be either the amplitude and phase shift, or the amplitudes of the in-phase and quadrature components, usually normalised to the primary field strength. Depending on the system, only one frequency or several frequencies with different penetration depths are used in order to investigate different regions of the ground.

Time-domain electromagnetic (TDEM) methods, also called transient EM (TEM) methods, use a rapid switch-off of the transmitter current. The receiver measures the signal during the time after the transmitter current has been switched off. Thus, there is no primary field, and the signal depends only on the conductivity of the subsurface. The typical shape is a rapid decay to zero, called a transient. Theoretically, one such decay curve would be sufficient (the decay curve is similar to the apparent resistivity sounding curve generated by VES surveys, see Chapters 1, 5, 7, 8 and 9), but the procedure is repeated several times and the signals are stacked in order to increase the signal-to-noise ratio.

2.2.2 Resolution and penetration depth

The resolution properties and penetration depth of an EM system depend on the frequency range or the time range over which measurements are taken, the geometry (e.g. the distance between transmitter and receiver) and the electrical conductivity. To characterise a system, it is useful to introduce the *skin depth*, given by

$$\delta = \sqrt{\frac{2\rho}{\omega\mu}} \approx 500\sqrt{\frac{\rho}{f}} \; [\text{m}], \qquad (2.1)$$

where ρ is the electrical resistivity, μ is the magnetic permeability, and $\omega = 2\pi f$, where f is the frequency. The second part of the equation is useful for a quick estimate, where ρ is given in $\Omega\,\text{m}$ and f in $1/\text{s}$. If the source is very far away from the receiver, the skin depth is the depth where the amplitude of the electromagnetic field has decayed to $1/e$ of the value at the surface, and is often used as a measure of the investigation depth. If a transmitter close to the receiver is used, the geometry has to be taken into account, which is done through the *induction number*,

$$I = \frac{r}{\delta}. \qquad (2.2)$$

Here, r characterises the dimension of a system, i.e. the distance between transmitter and receiver, or the size of a large transmitting loop. The induction number relates the size of the system to the skin depth. For a system transmitting at $f = 10\,\text{kHz}$, with a transmitter–receiver spacing of $40\,\text{m}$, over a halfspace with $100\,\Omega\,\text{m}$, the skin depth is $50\,\text{m}$, and the induction number is 0.8, i.e. the system size is approximately the same as the skin depth. Many systems (especially in periglacial environments) operate in the so-called *low-induction-number regime*, where $I \ll 1$. In that case, the resolution properties and the penetration depth are dominated by the geometry and the size of the system rather than by the frequency.

For TDEM systems (in the time domain), similar estimates can be made if the skin depth is replaced by the *diffusion depth*,

$$\delta = \sqrt{\frac{4t\rho}{\mu}},$$ (2.3)

where t is the time.

The choice of the frequency or time range, the sensor type and distance between transmitter and receiver leave room for many different EM systems with different properties. Compared to DC resistivity techniques, EM methods are more complicated, and the data depend on more parameters and are more sensitive to distortions. In order to avoid pitfalls and misinterpretation, a thorough knowledge of the theoretical background and experience is required. In the following, we will discuss in more detail two-coil frequency-domain systems, transient EM sounding and very-low-frequency/radiomagnetotelluric systems.

2.2.3 Frequency-domain two-coil systems

Principles

This section describes methods that use two small transmitter and receiver coils in the frequency domain designed for shallow investigation at high resolution (sometimes called Slingram methods; Spies and Frischknecht 1991). The coils often consist of simple portable loops, with a diameter of approximately 1 m, sometimes smaller coils with many windings and a ferrite core are mounted on a beam. These instruments can easily be carried by one person and measurements are taken while carrying the instrument over the ground at walking speed. Several different geometries between transmitter and receiver are in use. Figure 2.3 shows the most important ones and their established nomenclature, which refers to the plane of the loops, and not to the axes of the magnetic dipoles. The HCP (horizontal coplanar) and VCP (vertical coplanar) configuration are most common.

Some systems have been designed only for shallow conductivity mapping. They typically use the HCP or VCP configuration at a fixed distance of a few metres, and a single frequency, typically around 10 kHz. The instruments output a value for the apparent conductivity, obtained from the amplitude of the out-of-phase component normalised by the primary field. The primary field is assumed to be equal to the in-phase component. The contribution of the secondary field to the in-phase component can be neglected, as was illustrated in Figure 2.2. For low

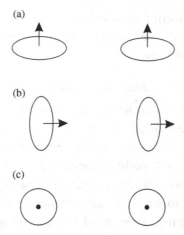

Figure 2.3. Nomenclature of common loop–loop configurations (after Frischknecht *et al.* 1991). (a) Horizontal coplanar (HCP), (b) vertical coplanar (VCP), (c) vertical co-axial (VCA).

induction numbers, and a homogeneous halfspace, the relationship between conductivity σ and the ratio of in-phase and out-of-phase components is

$$\sigma = \frac{4}{\omega\mu\,r^2}\frac{H_s}{H_p}, \tag{2.4}$$

where H_s and H_p are the secondary and primary magnetic fields, respectively. The conductivity value obtained in this way is called apparent conductivity, because it equals the true conductivity only in the case of a homogeneous halfspace. In general, the value will be a weighted average over the volume that contributes to the signal. Inversion or modelling techniques are required to derive a spatial conductivity distribution from the apparent conductivities measured at different spacings and locations (similar to the apparent resistivity in DC resistivity, see Chapter 1). Also note that a low induction number is the condition for the applicability of Equation (2.4). This may be violated in very conductive environments. A typical member of this instrument group is applied in Chapter 8, where a FEM system is used to measure horizontal conductivity variations in the uppermost 6 m at almost walking speed.

Another group of more flexible systems uses two separate loops that can be operated at several spacings and different frequencies. Spacings vary between 10 and 500 m, frequencies may range from 40 Hz to 56 kHz. Thus, the low-induction-number assumption is not always fulfilled, and the instruments provide both in-phase and out-of phase readings, typically normalised to the primary field and given in per cent or parts per million (ppm). Usually, at least HCP and VCP configurations can be used, and coil distances and corresponding frequencies can be chosen such that the induction number is the same for each frequency–distance combination. By this means, several measurements with different penetration depths are possible at each location.

Data acquisition

Data are almost invariably taken along one or several parallel straight profiles (see Chapter 8). Due to vegetation and other logistical constraints caused by the connecting cable between transmitter and receiver, the coils are usually in-line with the profile, although the broadside configuration may have advantages (e.g. better resolution of a target). When several coil spacings or configurations are used, it may be necessary to walk the same profile several times, although some equipment allows efficient measurement of several configurations at a single run. The spacing of stations along the profile or between lines depends on the desired resolution. As a rule of thumb, the maximum spacing between the loops should be at least twice the expected depth of the target, and the station spacing typically is 1/2 or 1/4 of the loop spacing.

Field procedures are dominated by the need for accurate positioning. Since the primary field varies with $1/r^3$, a 1% error in coil spacing will cause a 3% error in the reading. The coil orientation is generally less critical. When the HCP configuration is used and both coils are on the same slope, the coils should be oriented parallel to the surface. If large spacings are used and transmitter and receiver operator have no visual contact, preliminary work may be necessary to survey the exact distances, and coil angles need to be fixed based on topographic information and an inclinometer. It is also possible to apply corrections for wrong spacing or coil orientation to the in-phase and quadrature data, as described in Frischknecht *et al.* (1991).

Electromagnetic instruments can be sensitive to external influences. When coils are mounted on a beam, mechanical stress may change the relative position of the coils, which can cause systematic errors, and temperature changes can cause variation of the instrument properties (see below). Some manufacturers describe calibration procedures or simple checks that are based on the fact that the quadrature component should be zero in very resistive terrain or at sufficient distance from the ground. In the field, it will always be wise to observe the data quality critically and carry out repeated measurements at the same location to discover instrumental problems and remove a possible drift (Hauck *et al.* 2001).

Man-made noise is another factor that needs to be taken care of. Power lines generate an electromagnetic field that distorts the data and can be so strong that measurements are impossible. Small spacings will generally be less sensitive to power-line noise than large spacings. Otherwise, little can be done except adjusting the traverses and avoiding power lines wherever possible. Subsurface metallic conductors, which do not actively create their own EM field, cause systematic reproducible anomalies in the data. They may easily be identified if the station spacing is sufficiently dense; otherwise they appear as outliers or, in the worst case, they distort the data and cause misinterpretation.

Data processing

The first step in any data processing should be a careful visual inspection to identify and remove outliers, anomalies due to man-made conductors and instrumental problems. Corrections for spacing, coil misorientation and instrument calibration need to be carried out before data are interpreted. It may be useful to plot topography along with the data in order to identify correlations that may be due to geomorphology/geology or remaining spacing and alignment effects. If two-dimensional coverage is obtained, data are colour coded and displayed as a map (Figure 2.4).

In the simplest case, the profile displays or maps are interpreted directly by correlating lateral conductivity gradients with geological boundaries. Only in the case of known or assumed homogeneous geology can the data be interpreted in terms of frozen/unfrozen anomalies or subsurface water/ice occurrences. Further processing is possible to enhance features that are desired or to suppress undesired features. Taking the difference of data recorded at different spacings may reduce the effect of near-surface resistivity variations, whereas normalising apparent conductivity data with a background conductivity value or presenting them as relative conductivity variations (Figure 2.4) may simplify comparison with other data. If a specific target is investigated, the measured data may be compared with pre-calculated curves for certain models (e.g. Frischknecht *et al.* 1991).

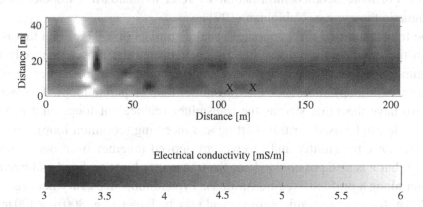

Figure 2.4. EM-31 conductivity map of a $200\,\text{m} \times 50\,\text{m}$ area at the permafrost station on the Schilthorn, Swiss Alps. The locations of two permafrost boreholes are marked with crosses. The highly conductive region in the eastern part of the area (between 15 and 30 m on the *x*-axis) corresponds to a buried metallic sewage pipe (modified after Hauck *et al.* 2001).

If the parameters of an approximately layered structure are important, a qualitative interpretation is not feasible. In that case, an inversion has to be carried out, where the layered model is automatically altered until the calculated data fit the measured data within a given error margin (similar to the inversion of vertical electrical soundings (VES) in DC resistivity surveys). Inversion codes usually parameterise the subsurface in terms of distinct horizontal layers with sharp boundaries and require a starting model to be entered. Note that some algorithms are based on the low induction number assumption. In this case, the contribution of each layer to the measured total response depends only on depth, and not on conductivity. This procedure will lead to misinterpretation if the low-induction-number assumption fails, which easily happens in conductive terrain.

2.2.4 Time-domain electromagnetic sounding

Principles

In time-domain methods, measurements are taken after a rapid switch-off of the transmitter current. The absence of the primary field at the time when the actual measurements are taken is generally considered an advantage, because the data are not dominated by a signal that does not carry information about the conductivity structure. Theoretically, there is equivalence between time-domain and frequency-domain methods in the sense that data can be transformed from one domain into the other. The switch-off signal carries the same information as a measurement of the full spectrum in the frequency domain. There are, however, many practical differences, a discussion of which is beyond the scope of this chapter. For more detailed information we refer to standard textbooks on electromagnetic theory, e.g. Nabighian (1991).

The transmitter of TEM soundings usually consists of a large loop that is laid out on the ground in a square. Figure 2.5 indicates the most commonly used configurations. The receiver may be a smaller coil placed in the centre of the transmitter (in-loop), or outside the transmitter loop (separate loop). The receiver can also have the same size as the transmitter (coincident loop), and even the same cable can be used for transmitting and receiving (common loop). For these configurations, transmitter and receiver are moved together from one station to the next, but it is also possible to keep the transmitter location fixed and move the receiver along a profile (fixed transmitter). Typical loop sizes vary between 20 m and 200 m for moving configurations, and may be larger (e.g. 800 m × 400 m) for fixed transmitter configurations. It is also possible to use a cable that is grounded at both ends (electric dipole transmitter) and transmitter–receiver spacings up to several kilometres. These systems are designed for large penetration depths of several kilometres and will not be discussed here.

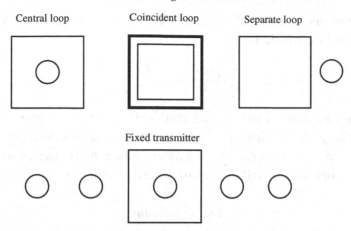

Figure 2.5. Common transmitter–receiver configurations for TEM soundings (redrawn after Greinwald *et al.* 1997). The large squares indicate the transmitter loop, the circle or smaller square are receivers.

For moving transmitter configurations, the loop diameter is the key parameter to characterising the size of the system, similar to the transmitter–receiver spacing for the two-coil FEM systems. The investigation depth is a function of geometry, conductivity, transmitter moment, noise level and time range, but in principle penetration depths can be achieved that are much larger than the loop diameter. On the other hand, very shallow investigation is not possible with TEM systems, partly because the switch-off of the transmitter current is not instantaneous, but uses a finite ramp time, which sets a threshold on the earliest time channels that can be measured. In general, TEM sounding methods are used when the very shallow depths are not important and a large investigation depth is desired, whereas two-coil FEM systems are preferred for near-surface, high-resolution investigation.

TEM soundings are categorised to take place in the *near zone*, *far zone*, or *intermediate zone*, depending on the transmitter size and the time channels. The near zone is defined by

$$\frac{2t}{\sigma\mu a^2} > 5, \tag{2.5}$$

where a is the diameter of a circular loop, which is mathematically easier to treat than a square. If a square loop is used in practice, a will be the radius of a circular disc of equal area. The near zone corresponds to low induction numbers, and the latest time channels of in-loop measurements are almost always in the near zone.

The late-time asymptotic voltage of the in-loop over a homogeneous halfspace can be written in a simple form:

$$V(t) = -\mu A \frac{\partial H_z}{\partial t} = \mu A \frac{I \sigma^{3/2} \mu^{3/2} a^2}{20 \pi^{1/2} t^{5/2}}, \qquad (2.6)$$

where A is the moment of the receiver coil, and I is the transmitter current. The late-time voltage decays with $t^{-2.5}$ and increases with conductivity. Equation (2.6) may be used to calculate the conductivity of a halfspace, or to define an apparent late-time conductivity or resistivity by inverting for σ.

Data acquisition

Before the actual survey, the optimum transmitter loop size should be determined. As a rule of thumb, the loop size should be approximately equal to the assumed target depth. It is recommended to try different loop sizes for the first station and determine which one gives the best results. Typical sizes often used are $50\,\text{m} \times 50\,\text{m}$ or $100\,\text{m} \times 100\,\text{m}$ (see Chapter 7). For small loop sizes ($<25\,\text{m}$) it may be difficult to achieve good data quality. The station spacing is typically chosen to be equal to the loop size (no overlap) or half the loop size (50% overlap). For quick reconnaissance surveys in areas where the lateral variations are known to be smooth, this spacing may be even larger. The time channels (or time gates), i.e. the number and sequence of the single measurements within one sounding, are usually fixed for each instrument. Some manufacturers offer different time ranges, but in general they are between $7\,\mu s$ and $100\,\text{ms}$. If there is a choice, the diffusion depth (Equation (2.3)) based on assumed conductivities and target depth may be used to support the decision.

It is recommended to survey the corners of the loop, although errors due to geometry are much less important than for FEM methods, because there is no primary field. Since a movement of a loop in the earth's magnetic field induces a voltage that will distort the data, wind may be a problem. To reduce these effects, all cables should be placed as firmly on the flat ground as possible. As for FEM methods, power lines generate EM noise, and metal conductors in the ground will generate undesired anomalies. They should be avoided or recorded in the measurement protocol so that they can be considered during interpretation.

Data processing

For most available systems, the stacking, i.e. the averaging of repetitive signals to reduce sporadic noise, and filtering, e.g. to suppress power-line noise, is carried out during data acquisition. The output is a curve of voltage versus time or late-time apparent resistivity versus time, including error bars calculated from the stacking procedure. The data should be plotted on a log–log scale for quality control. In general, the curve should be smooth, with no rapid changes. For the in-loop, common-loop and coincident-loop configurations, the transient does not change sign over a layered halfspace. Therefore, sign changes, which appear as a sharp notch when plotting absolute values on a log–log scale, are an indication of strong lateral conductivity gradients. For the separate-loop configuration, sign changes occur even over a homogeneous halfspace, and are not indicative of any particular structure. Profile plots, i.e. plotting data along the profile for a fixed time, are useful to identify local distortions and obtain an impression of the spatial variability.

The standard interpretation method is one-dimensional (1D) inversion, which uses the same principle as in 1D inversion of vertical electrical soundings

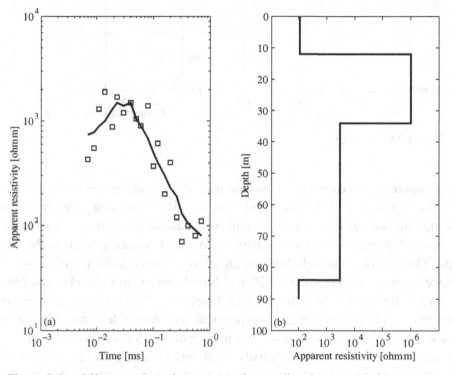

Figure 2.6.   Offset transient electromagnetic sounding data recorded across (a), (b) rock glacier Muragl, Swiss Alps, and (c), (d) rock glacier Murtel, Swiss Alps. (a), (c) Measured (squares) and calculated (solid line) TEM sounding data, represented as late-time apparent resistivities obtained from the voltages using Equation (2.6). (b), (d) Layered inversion model calculated from the apparent resistivity data set shown on the left. Depth of rock glacier base at the Murtel borehole is marked by the horizontal line. In both cases the high-resistivity zones correspond to massive ground ice occurrences in the rock glacier core.

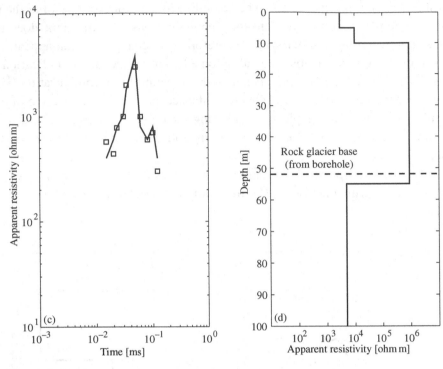

Figure 2.6.   (cont.)

(see Chapter 1). A layered model is automatically found such that the calculated data for this model fit the measured data as well as possible. No simplifying assumptions such as low induction numbers are made; the full equations have to be solved. Figure 2.6 shows a typical example from two rock glaciers in the Swiss Alps. The inversion models indicate an approximately 5–10 m thick unfrozen surface layer ($\rho < 1$ kΩ m) and a 20 m (Muragl) and 40 m (Murtel) thick frozen layer ($\rho > 500$ kΩ m) above the bedrock layer ($\rho < 10$ kΩ m), which is in good agreement with complementary geophysical and borehole data (Maurer and Hauck 2007). Care must be taken when interpreting the 1D inversion models. Different models may yield an equally good data fit, and a-priori information or the analysis of several stations along the profile is necessary to decide which model is the best description of the true situation. Some layers may not be resolved, i.e. their thickness or resistivities may be changed without changing the data fit. In that case, no conclusions can be drawn about that particular layer. Typically, for thin layers it is only possible to determine the product of thickness and conductivity. Most commercially available inversion packages offer tools to analyse this ambiguity. In any case, it is a simple matter to carry out a sensitivity analysis where layer parameters are changed in a systematic way to evaluate the data resolution.

Although TEM soundings have their maximum sensitivity vertically below the station, the data may be affected by lateral conductivity variations, called 3D effects. The assumption of horizontal layering may then lead to a misinterpretation. Sometimes these effects can easily be identified, for example by a sign reversal in the in-loop data. Another typical 3D feature is a steeply increasing slope of the late-time apparent resistivities, which leads to an additional, anomalously good conductor when inverted with a 1D model. There are no general rules to identify 3D effects, but a smooth variation of the models along a profile is one criterion that should be fulfilled.

2.2.5 VLF and radiomagnetotellurics

Principles

The VLF (very-low-frequency) method uses existing transmitters in the frequency range between 10 kHz and 30 kHz, which are normally used for marine communication and exist all over the world. At a large distance from the transmitter, the electromagnetic field can be considered as a polarised plane wave with horizontal electric and magnetic field components. The electric field points towards or away from the transmitter, the magnetic field is perpendicular to the transmitter (Figure 2.7). In its original form, VLF measures the ratio of vertical to horizontal magnetic field amplitudes. The vertical magnetic field is caused by induction in confined bodies or horizontal conductivity gradients and vanishes over homogeneous or horizontally layered ground. Therefore, VLF is sensitive to lateral conductivity variations only. The secondary field (see also Section 2.2.1) caused by lateral conductivity gradients has both a vertical and a horizontal component. However, the primary horizontal field is much larger than the secondary field, and thus the horizontal field can be assumed to consist of the primary field only, whereas the vertical component consists of the secondary field only, i.e.

$$H_y \approx H_y^p,$$
$$H_z \approx H_z^s. \tag{2.7}$$

As explained in Section 2.2.1, the electromagnetic fields may be separated into a component that is in-phase with the primary field, and a component that is 90 degrees out of phase, and thus the measured VLF data consist of

$$\text{In-phase}\left(\frac{H_z}{H_y}\right) \quad \text{and} \quad \text{Quadrature}\left(\frac{H_z}{H_y}\right).$$

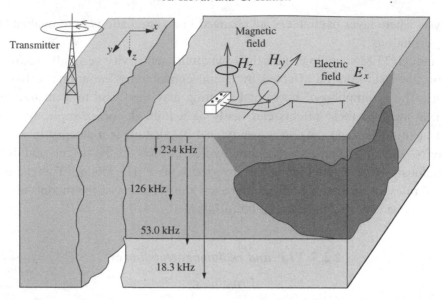

Figure 2.7. Schematic sketch of VLF and RMT methods (modified from Zacher *et al*. 1996). VLF measures the ratio of vertical to horizontal magnetic fields. VLF-R and RMT measure the ratio of horizontal electric and magnetic fields. RMT uses a broad frequency range, with penetration increasing with decreasing frequency.

For a more detailed treatment of the theoretical background and practical aspects of the VLF method we refer to Paterson and Rönkä (1971), McNeill (1990), McNeill and Labson (1991) and Nabighian (1991).

In order to derive quantitative conductivity information, electric fields need to be measured. This method is usually called VLF-R, where the 'R' stands for resistivity. The apparent resistivity is calculated from the ratio between electric and magnetic field components (Figure 2.7), which is called impedance and is denoted Z:

$$\rho_{xy}(\omega) = \frac{1}{\omega\mu_0}\left|\frac{E_x}{H_y}\right|^2 = \frac{1}{\omega\mu_0}\left|Z_{xy}\right|^2. \qquad (2.8)$$

The phase shift between electric and magnetic fields is also measured and used during the interpretation:

$$\varphi_{xy}(\omega) = \tan^{-1}\left(\frac{\mathrm{Im}\left(Z_{xy}\right)}{\mathrm{Re}\left(Z_{xy}\right)}\right). \qquad (2.9)$$

The VLF and VLF-R methods usually measure at a single frequency only, and thus are used for mapping or profiling. In order to obtain a vertical resolution, measurements need to be taken over a broader frequency range. The radiomagnetotellurics (RMT) method is an extension of the VLF-R method towards higher frequencies. Typically, data are measured at four frequencies in the range between 10 kHz and 300 kHz. The high-frequency signals originate from broadcasting or time-signal transmitters. The use of different frequencies gives different penetration depths, which may be estimated using the skin depth given in Equation (2.1). Radiomagnetotellurics is a relatively novel technique that has been successfully used in environmental applications (e.g. Zacher *et al.* 1996, Tezkan 1999, Pellerin and Alumbaugh 1997). Recently, a new system was introduced that is able to measure the full impedance tensor, which is required in the case of 3D resistivity distribution or unknown strike direction (Pedersen *et al.* 2005).

Data acquisition

VLF measurements are almost invariably made along profiles. Usually, a geological strike direction is known or assumed beforehand. Ideally, a transmitter is found that is located in the strike direction, producing a primary magnetic field perpendicular to the strike. The profiles are then aligned across the strike, i.e. in the direction of the magnetic field. The direction of a transmitter is found by searching for a maximum (or minimum) of the magnetic field, and usually a sufficient number of transmitters are available to find one with a suitable azimuth. The station spacing typically varies between 1 and 10 m.

When electric fields are measured (VLF-R and RMT), the profiles also run across a known or assumed strike direction. However, in this case the transmitter direction may be chosen either along or across the strike. The data recorded with the magnetic field parallel to the strike are called the TM-mode, and the perpendicular components are called the TE-mode. The two modes have different sensitivities with respect to the conductivity structure, a discussion of which is beyond the scope of this introduction (see Berdichevsky *et al.* 1998 and references therein). If possible, recording both modes and combining the data during the interpretation is advisable. This involves searching two different transmitters at approximately perpendicular directions for a pair of nearby frequencies. The measurements can be carried out by one person, but since two electrodes, at typical distances of 5–10 m, have to be put into the ground, a second person will increase efficiency.

Data processing

VLF data are usually plotted along the profiles or contoured if several parallel profiles have been measured. However, the anomalies are often complex and the interpretation is not so straightforward. For example, over a confined conductive body, the data show a zero crossing over the centre of the body, instead of a

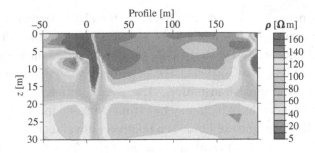

Figure 2.8.   Resistivity section obtained from a 2D inversion of RMT data (from Hördt *et al.* 2000). For colour version see Plate 2.

maximum or minimum. Therefore, filtering techniques based on the subsurface current distribution have been suggested (Fraser 1969, Karous and Hjelt 1983). The processed data show a maximum over the conductivity anomaly, which facilitates the interpretation. Since the processing may also generate artefacts, it is always advisable to look at both filtered and raw data in parallel.

Quantitative inversion techniques have also been discussed (e.g. Kaikkonen and Sharma 2001). Recently, Becken and Pedersen (2003) developed a scheme to calculate apparent resistivity maps from VLF data, and Pedersen and Becken (2005) suggested a new filtering technique to calculate vertical sections of conductivity from VLF data.

For RMT data, 2D inversion is a standard interpretation tool. Just like multi-electrode resistivity data (Chapter 1) a 2D model may be calculated from the multi-frequency data recorded along a profile. Figure 2.8 (Plate 2) shows an example from an industrial waste site. The waste is characterised by low resistivities (red colours) within a more resistive background. The RMT method is recognised as a promising tool in groundwater and environmental investigations. Particular issues arising from the application in rough terrain have been discussed by Hördt and Zacher (2003), and one example will be given in the following section.

2.3 Periglacial applications and particularities

The aforementioned methods have been applied in a number of glacial and periglacial studies, but mostly in arctic regions and Antarctica (e.g. Hoekstra and McNeill 1973, Sartorelli and French 1982, Rozenberg *et al.* 1985, Scott *et al.* 1990 and references therein, Todd and Dallimore 1998, Harada *et al.* 2000 and 2006, Ingeman-Nielsen 2005, Cockx *et al.* 2006, Yoshikawa *et al.* 2006). Only a few studies show examples of successful applications of EM methods in mountainous regions, some of which are discussed in Hauck and Vonder Mühll (2003). FEM methods are used by Schmöller and Frühwirth (1996) to detect ground ice in a rock glacier, by Hauck and Vonder Mühll (1999) and Cannone *et al.* (2003) to map shallow ground ice occurrences in the European Alps and by Hauck *et al.* (2004) to detect the altitudinal limit of a mountain permafrost occurrence in Norway. Hauck *et al.* (2001) compare different FEM and TEM methods for the application in mountain permafrost studies. Beylich *et al.* (2003) use RMT to determine the regolith thickness at a periglacial drainage basin in Swedish Lapland, Hauck *et al.* (2003) compare TEM soundings on an Alpine rock glacier with corresponding ERT measurements, and Bucki *et al.* (2004) combine several TEM soundings to map ground ice variations in a rock glacier in the Rocky Mountains. Case studies using FEM and TEM are also presented in Chapters 7 and 8.

As mentioned above, the response of a FEM or TEM instrument depends not only on the instrument specifications and the electrical properties of the sub-surface, but may be influenced in a number of unwanted ways, e.g. changing surface characteristics, instrument height over ground, electrical power lines, atmospheric lightning in the vicinity or instrument drift. The response of metallic objects, power lines or lightning in an otherwise extremely resistive environment is very pronounced and can usually be distinguished from the subsurface target signal, even though it may prohibit data acquisition. The influences of differences in instrument height, a changing snow cover or instrument drift are much smaller and can easily be mistaken for a subsurface target signal. In the following, some of the particularities of electromagnetic applications in mountainous and arctic terrain are described.

2.3.1 Low conductivities, low-induction-number regime and penetration depth

One of the characteristics of periglacial and glacial environments is the highly resistive surface and subsurface conditions. An exception is surveys in arctic lowland regions in summer, where conditions are more conductive due to the high unfrozen water content, the large fraction of organic material and the saline conditions in coastal regions (see e.g. Chapter 7). Observed resistivities in the Arctic are around 1–10 kΩ m, whereas typical mountain permafrost values range between 10 kΩ m and several MΩ m (see Table A.2). The electromagnetic response of these highly resistive targets is therefore close to the resolution limit of most EM instruments, which may be a reason why these systems are not as popular in mountain terrain compared to arctic environments.

Due to the low conductivities observed in alpine periglacial terrain, the low-induction-number approximation (Equation (2.2)) is almost always valid – which facilitates data processing, as the apparent conductivity of the ground is simply proportional to the ratio between the secondary and the primary magnetic field (Equation (2.4)). However, the magnetic field will decay very rapidly in a highly resistive subsurface resulting in small signal strengths at greater depths. Harada *et al.* (2000 and 2006) report penetration depths of 50 m in Alaska and Mongolia and around 400 m in Siberia. Todd and Dallimore (1998) even report penetration depths of >500 m in the Mackenzie Delta.

2.3.2 Frequency dependence of resistivity

In addition to the frequency dependence of the signal due to the skin effect described by Equation (2.1), the resistivity of the rock itself may also display a dependence on frequency. Electromagnetic energy transport may be divided into high- and low-frequency domains. At higher frequencies the influence of the dielectric constant ε dominates and energy is transported *wave-like* (used in ground-penetrating radar (GPR) surveys, see Chapter 4). Corresponding observed resistivity values decrease with increasing frequency. At lower frequencies, electrical conductivity and measurement geometry dominate and energy transport is *diffusive* (used in direct current (DC) resistivity and FEM methods). Near the so-called DC limit (frequency 0) resistivity is determined by ionic conduction in the pore fluid and is independent of frequency.

For typical earth material the transition between frequency-dependent and frequency-independent behaviour occurs near 10 MHz. For very resistive material this transition is lowered to frequencies around 1 kHz (Onsager and Runnels 1969, Grimm 2002), which is well within the frequency range of most common FEM instruments. This transition from dominantly conduction-loss to dielectric-loss mechanisms (from frequency-independent to frequency-dependent resistivities) complicates the comparison and interpretation of different electrical and electromagnetic field survey results on permafrost (Olhoeft 1978, Kneisel and Hauck 2003, Hauck and Kneisel 2006). This is one of the reasons why resistivity values obtained from DC vertical electrical soundings and EM soundings may differ substantially (see Chapter 7). In addition, the different penetration depths of sensors with different operating frequencies may further complicate the comparison of survey results of different electrical and electromagnetic instruments. Hauck *et al.* (2001) and Geonics (undated) discuss the suitability of multi-frequency EM instruments for application in resistive terrain. Persson and Pedersen (2002) investigate the effect of frequency-dependent resistivity specifically for the RMT method.

Figure 2.9 shows an example of RMT data recorded in Jotunheimen, Norway. The aim of the survey was to image lateral variation of permafrost distribution. The profile was initially investigated with electrical resistivity tomography, FEM

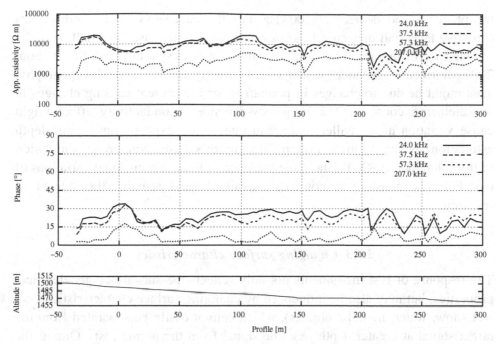

Figure 2.9. Radiomagnetotelluric data (TE-mode) recorded in Jotunheimen, Norway. Top panel: Apparent resistivity along the profile for four different frequencies. The 57.3 kHz transmitter replaces the 37.5 kHz transmitter at 100 m. Middle panel: Phase data. Bottom panel: Topography along the profile.

and refraction seismics (Isaksen *et al.* 2002, Hauck *et al.* 2004). The RMT data have reasonable quality and agree well with the previous geophysical surveys, but they also illustrate several of the peculiarities discussed above. The apparent resistivity curves (top panel) show high resistivities between 1 and 20 kΩ m. The resistivity range of this particular instrument is limited to 20 kΩ m, which is why some of the lowest frequency data are clipped at this value. The skin depth of the highest frequency is approximately 70 m, indicating that the penetration depth is higher, and resolution is much lower, compared to DC resistivity measurements. Another typical feature of the RMT measurements is the change of transmitters (frequencies) at 100 m for the intermediate frequency, which was necessary because the transmitter used at the beginning (37.5 kHz) stopped transmitting during the survey.

The phase data (middle panel) are almost all below 30 degrees, some below 15 degrees. This is remarkably small compared to the values over a homogeneous halfspace (45 degrees). Small phase values may be caused either by a highly resistive layer beneath a more conductive one, or by dielectric loss mechanisms. The work by Persson and Pedersen (2002) and additional modelling results not shown here indicate that the frequency dependence of resistivity can easily lower the phases by 30 degrees at 200 kHz. For the data shown here, both effects (resistive layer and dielectric loss) probably add up to cause the small phases.

One specific feature of the data set is the change in spatial variability at about 200 m horizontal distance, which is not directly correlated with topography. This might be due to changes in permafrost or ice content causing changes in the dielectric constant. The frequency-dependent conductivity effect might cause variation at a smaller scale than one would expect from a skin depth estimation, which is greater than 70 m in this case. Numerical simulation of the behaviour of RMT data in the presence of three-dimensional variations of the dielectric constant is not easily available and remains the subject of research.

2.3.3 Changing surface characteristics

The response of EM instruments not only reflects the subsurface geoelectrical properties, but may also be influenced by changing surface characteristics (ice, wet snow, water, metallic objects), which cannot easily be separated from the target signal at greater depth (e.g. the signal from the permafrost). One of the major advantages of EM survey methods as compared to their electrical counterparts (VES, ERT) is the absence of the necessity to establish direct electrical (galvanic) contact between the instrument sensor and the ground. This characteristic facilitates the applicability of EM methods on debris-covered mountain surfaces and on snow-covered terrain, where electrical methods are often difficult to apply. However, the presence of a snow cover introduces an additional electrically relevant layer, which has to be taken into account for the interpretation of EM data sets. Two properties of the snow cover are thereby important: (1) the thickness and (2) the unfrozen water content, as well as their spatial variability.

Figure 2.10 shows an example of a comparison of two FEM instruments near the Stelvio Pass, Italy, where the aim was to map possible permafrost occurrences. The results from both instruments are very similar, with a dominant conductivity maximum observed between stations 150 and 200 m. This coincided with a wet snow patch on the surface in the middle of an otherwise snow-free survey line. Since unfrozen water has a much higher conductivity than the underlying material, this wet snow patch masks any possible underlying permafrost, as the conductivity signal of the latter is much smaller. If the survey is conducted on snow-covered ground, care must be taken to ensure that the snow cover is homogeneous with respect to water content and thickness, because these heterogeneities may be misinterpreted as thermal (permafrost) or geological variations.

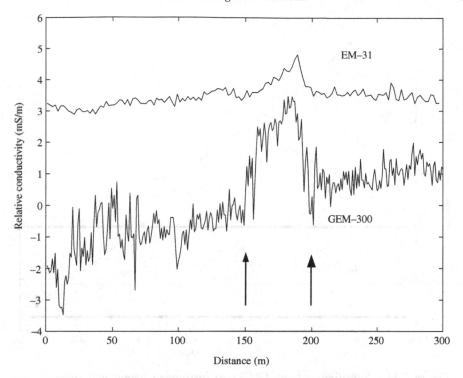

Figure 2.10. Comparison between Geonics EM-31 (9800 Hz) and Geometrics GEM-300 (19975 Hz) near the Stelvio Pass. The black arrows mark the beginning and end of the wet snow patch. The response of both instruments is similar: a sharp increase in conductivity due to the increase in unfrozen water (after Hauck *et al.* 2001).

The conductivity response also depends on instrument height, since the measured apparent conductivity is a function of all the specific conductivities in a subsurface column beneath the instrument, including the air and snow layers between the instrument and the ground surface. Changing air-layer and snow-cover thicknesses will change the measured apparent conductivities even where the ground conductivity is uniform (Figure 2.11).

2.3.4 Drift

One of the biggest problems encountered during EM surveys in mountainous terrain is the instrument drift. Ground conductivity meters are generally designed in such a way that the shift of the zero level with time, temperature etc. is within about 1 mS/m (McNeill 1990). The accuracy may vary between instruments and can depend on age, battery charge, sensor temperature and surface temperature of the instrument. The accuracy of the zero level can be tested in a drift experiment

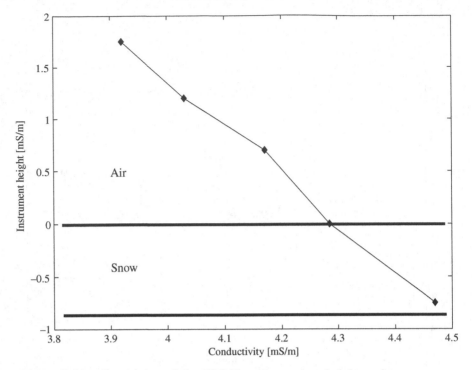

Figure 2.11. Sensitivity of the EM-31 to instrument height and snow-cover depth using a data set from the Schilthorn, Swiss Alps (after Hauck *et al.* 2001). As air and dry snow have smaller conductivities than the ground, the total conductivity increases with decreasing instrument height. This is only true for dry snow as increasing water content in wet snow would lead to an increase in total conductivity.

where the instrument is placed on the ground and measurements are taken continuously for up to the expected survey duration. Figure 2.12 shows an example from a 30 minute drift experiment showing a marked increase in conductivity (0.3 mS/m) during the first 15 minutes, followed by smaller fluctuations. For most geophysical surveys this accuracy is sufficient as the target signal is usually much larger than the drift amplitude. However, in permafrost studies the permafrost signal can be as small as 0.4 mS/m (Hauck and Vonder Mühll 1999) so that this instrumental drift may cause severe problems. In combination with the shift of the zero level in most conductivity meters, McNeill (1980) concluded that accurate absolute conductivity values are difficult to obtain in very resistive ground. However, relative changes in conductivity can be accurately estimated. Therefore, results are often presented as relative conductivity variations against a background conductivity value.

The problem of instrument drift can increase when temperature changes add to the natural drift of the instrument. If, for instance, the survey is initially

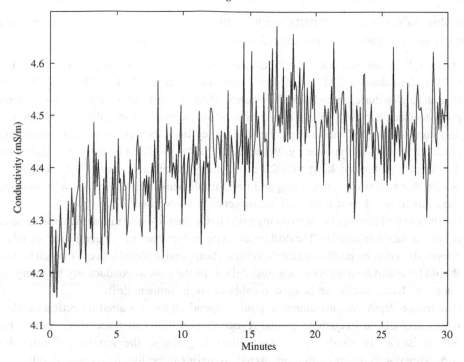

Figure 2.12.   Drift experiment with the Geonics EM-31. Measurement duration was 30 minutes, where the instrument was set on the ground and conductivities were measured every 5 seconds (after Hauck *et al.* 2001).

conducted over a cold snow surface on a day with increasing solar radiation during the day, instrument drift may become more pronounced the longer the survey continues due to increasing temperature gradients (e.g. Hauck *et al.* 2001). Because the drift rate of the instrument is not constant but depends on external variables such as temperature, filtering for drift by applying a constant average rate of change may not overcome this problem. It is concluded that repeated measurements at fixed locations throughout a survey are necessary to quantify instrument drift, and the problem is likely to be worse when radiation or air temperatures are changing during the survey.

2.4 Conclusions

An introduction to various electromagnetic induction methods as well as an analysis of the applicability and reliability of some of these methods for peri-glacial environments have been presented. Particularities of periglacial environments may include the low electrical conductivities observed, the presence of a (possibly) inhomogeneous snow cover, strong temperature gradients leading to enhanced instrument drift and strongly heterogeneous surface conditions. The possible influence of these particularities on the survey results was discussed for a variety of applications in periglacial research. Key results include:

- Most EM instruments can be differentiated into three groups: frequency-domain EM (FEM), time-domain EM (TEM) and very-low-frequency EM (VLF) including the radiomagnetotelluric method (RMT). Whereas FEM and VLF are mostly used for mapping purposes (see Chapter 8), TEM is usually applied for deep vertical soundings (see Chapter 7); they can also be conducted along horizontal survey lines, yielding quasi-two-dimensional conductivity distributions, see e.g. Bucki *et al.* (2004). RMT can be measured along two-dimensional profiles, yielding fully 2D conductivity distributions, as in ERT surveys.
- All FEM methods exhibit a strong and undesirable sensitivity to wet snow conditions. Care has to be taken if a laterally heterogeneous snow cover is present.
- Instrument drift can be a problem owing to the low signal-to-noise ratios often encountered in periglacial environments. The drift becomes more important the longer the survey takes. Survey lines may be partitioned into several sections, and repeated surveys of each section should be undertaken without long time delays. In the case of conductivity mapping, tie lines with base stations can be used to calibrate for instrument drift.
- Penetration depth and instrument response depend on the so-called induction number, which depends on frequency, ground resistivity and transmitter–receiver spacing. For high-resistivity materials, as on ice-rich rock glaciers, the low-induction-number approximation is always valid, but signal strength can be low for greater depths.

2.5 Checklist

Before starting a survey:

- Is the method used adequate to answer the scientific question?
- Are there any other useful complementary methods?
- What is the minimum expected number of profiles/soundings?
- What is the required line spacing for mapping surveys and how does this relate to the spatial extent of expected anomalies? What is the expected time required for the planned survey?
- Is there any information on the expected subsurface condition based on geomorphological and geological setting and interpretation?
- Is there any information about expected depth of anomalies?
- Is there any possibility of hidden metallic objects, cables etc.?
- Are there any other critical factors (power lines, other EM noise, surface heterogeneities, water, snow etc.) influencing the EM signal?
- Can the weather conditions influence the EM instrument outputs (temperature/radiation changes)?

During the survey:

- Mark carefully the edges of the survey grid and the locations of the proposed survey line (for FEM mapping).

- Conduct a drift experiment for EM instruments.
- TEM soundings: stack individual sounding curves as appropriate.
- Make backup copies of output data at regular intervals.

After the survey:

- Use postprocessing software that offers different inversion tools (e.g. smooth inversion *and* layered inversion).
- Make (whenever possible) cross-checks using complementary geophysical methods.
- Try to avoid the misinterpretation of critical data sets and repeat the survey if necessary.

REFERENCES

Becken, M. and Pedersen, L. B. (2003). Transformation of VLF anomaly maps into apparent resistivity and phases. *Geophysics*, **68**, 497–505.

Berdichevsky, M. N., Dmitriev, V. I., and Pozdnjakova, E. E. (1998). On two-dimensional interpretation of magnetotelluric soundings. *Geophysical Journal International*, **133**, 585–606.

Beylich, A., Kolstrup, E., Linde, N., Pedersen, L., Thyrsted, T., Gintz, D. and Dynesius, L. (2003). Assessment of chemical denudation rates using hydrological measurements, water chemistry analysis and electromagnetic geophysical data. *Permafrost and Periglacial Processes*, **14**, 387–397.

Bucki, A. K., Echelmeyer, K. A. and MacInnes, S. (2004). The thickness and internal structure of Fireweed rock glacier, Alaska, U.S.A., as determined by geophysical methods. *Journal of Glaciology*, **50**(168), 67–75.

Cannone, N., Guglielmin, M., Hauck, C. and Vonder Mühll, D. (2003). The impact of recent glacier fluctuation and human activities on the permafrost distribution: a case study from Stelvio Pass (Italian Central-Eastern Alps). *Proceedings of the 8th International Conference on Permafrost*, Zürich, Switzerland, 125–130.

Cockx, L., Ghysels, G., Van Meirvenne, M. and Heyse, I. (2006). Prospecting frost-wedge pseudomorphs and their polygonal network using the electromagnetic induction sensor EM38DD. *Permafrost and Periglacial Processes*, **17**, 163–168.

Fraser, D. C. (1969). Contouring of VLF-EM data. *Geophysics*, **34**, 958–967.

Fraser, D. C. (1978). Resistivity mapping with an airborne multicoil electromagnetic system. *Geophysics*, **43**(1), 144–172.

Frischknecht, F. C., Labson, V. F., Spies, B. R. and Anderson, W. L. (1991). Profiling methods using small sources. In *Electromagnetic Methods in Applied Geophysics*, ed. Nabighian, M. N., Vol. 2, Society of Exploration Geophysics, Tulsa, pp. 5–45.

Geonics (undated). *Application of Dipole–Dipole Electromagnetic Systems for Geological Depth Sounding*. Technical Note TN-31, Geonics Ltd.

Greinwald, S., Illich, B., and Schaumann, G. (1997). Transientelektromagnetik. In *Handbuch zur Erkundung des Untergrundes von Deponien und Altlasten, Band 3 – Geophysik*, eds. Knödel, K., Krummel, H. and Lange, G., Springer.

Grimm, R. E. (2002). Low-frequency electromagnetic exploration for groundwater on Mars. *Journal of Geophysical Research*, **107**, 2001JE001504.

Haas, C. (2004). Late-summer sea ice thickness variability in the arctic transpolar drift 1991–2001 derived from ground-based electromagnetic sounding. *Geophysical Research Letters*, **31**, L09402.

Harada, K., Wada, K. and Fukuda, M. (2000). Permafrost mapping by transient electromagnetic method. *Permafrost and Periglacial Processes*, **11**, 71–84.

Harada, K., Wada, K., Sueyoshi, T., Fukuda, M. (2006). Resistivity structures in alas areas in Central Yakutia, Siberia, and the interpretation of permafrost history. *Permafrost and Periglacial Processes*, **17**(2), 105–118.

Hauck, C. and Kneisel, C. (2006). Application of capacitively-coupled and DC electrical resistivity imaging for mountain permafrost studies. *Permafrost and Periglacial Processes*, **17**(2), 169–177.

Hauck, C. and Vonder Mühll, D. (1999). Detecting alpine permafrost using electromagnetic methods. In *Advances in Cold Regions Thermal Engineering and Sciences*, eds. Hutter, K., Wang, Y. and Beer, H., Springer, pp. 475–482.

Hauck, C. and Vonder Mühll, D. (2003). Evaluation of geophysical techniques for application in mountain permafrost studies. *Zeitschrift für Geomorphologie, Supplement*, **132**, 161–190.

Hauck, C., Guglielmin, M., Isaksen, K. and Vonder Mühll, D. (2001). Applicability of frequency- and time-domain electromagnetic methods for mountain permafrost studies. *Permafrost and Periglacial Processes*, **12**(1), 39–52.

Hauck, C., Vonder Mühll, D. and Maurer, H. (2003). Using DC resistivity tomography to detect and characterise mountain permafrost. *Geophysical Prospecting*, **51**, 273–284.

Hauck, C., Isaksen, K., Vonder Mühll, D. and Sollid, J. L. (2004). Geophysical surveys designed to delineate the altitudinal limit of mountain permafrost: an example from Jotunheimen, Norway. *Permafrost and Periglacial Processes*, **15**(3), 191–205.

Hoekstra, P. and McNeill, D. (1973). Electromagnetic probing of permafrost. *Proceedings of the 2nd International Conference on Permafrost*, Yakutsk, Siberia, 517–526.

Hoekstra, P., Sellmann, P. V., Delaney, A. (1975). Ground and airborne resistivity surveys of permafrost near Fairbanks, Alaska. *Geophysics*, **40**, 641–656.

Hördt, A. and Zacher, G. (2003). The radiomagnetotellurics method and its potential application in geomorphology. *Zeitschrift für Geomorphologie, Supplement*, **132**, 123–143.

Hördt, A., Greinwald, S., Schaumann, S., Tezkan, B. and Hoheisel, A. (2000). Joint 3D interpretation of radiomagnetotelluric (RMT) and transient electromagnetic (TEM) data from an industrial waste deposit in Mellendorf, Germany. *European Journal of Environmental and Engineering Geophysics*, **4**, 151–170.

Ingeman-Nielsen, T. (2005). *Geophysical Techniques Applied to Permafrost Investigations in Greenland*. Arctic Technology Centre, BYG·DTU, Technical University of Denmark, 177pp.

Isaksen, K., Hauck, C., Gudevang, E., Ødegård, R. S. and Sollid, J. L. (2002). Mountain permafrost distribution in Dovrefjell and Jotunheimen, southern Norway, based on BTS and DC resistivity tomography data. *Norwegian Journal of Geography*, **56**, 122–136.

Kaikkonen, P. and Sharma, S. P. (2001). A comparison of performances of linearized and global nonlinear 2-D inversions of VLF and VLF-R electromagnetic data. *Geophysics*, **66**, 462–475.

Karous, M. and Hjelt, S. E. (1983). Linear filtering of very-low-frequency (VLF) dip angle measurements. *Geophysical Prospecting*, **31**, 782–794.

King, M. S., Zimmerman, R. W. and Corwin, R. F. (1988). Seismic and electrical properties of unconsolidated permafrost. *Geophysical Prospecting*, **36**, 349–364.

Kneisel, C. and Hauck, C. (2003). Multi-method geophysical investigation of a sporadic permafrost occurrence. *Zeitschrift für Geomorphologie, Supplement*, **132**, 145–159.

McNeill, J. D. (1980). *Electromagnetic Terrain Conductivity Measurements at Low Induction Numbers*. Technical Note TN-6, Geonics Ltd.

McNeill J. D. (1990). Use of electromagnetic methods for groundwater studies. In *Geotechnical and Environmental Geophysics*, ed. Ward, S. H., Society of Exploration Geophysics, Tulsa, Vol. I, pp. 191–218.

McNeill, J. D. and Labson, V. F (1991). Geological mapping using VLF radio fields. In *Electromagnetic Methods in Applied Geophysics, Vol. 2: Application, Part B*, ed. Nabighian, M., Society of Exploration Geophysics, Tulsa, pp. 521–640.

Maurer, H. and Hauck, C. (2007). Geophysical imaging of alpine rock glaciers. *Journal of Glaciology*, **53**(180), 110–120.

Militzer, H. and Weber, F. (1985). *Angewandte Geophysik*, vol. 32, Springer.

Nabighian, M. N. (ed.) (1991). *Electromagnetic Methods in Applied Geophysics, Vol. 2*, Society of Exploration Geophysics, Tulsa.

Olhoeft, G. R. (1978). Electrical properties of permafrost. *Proceedings of the 3rd International Conference on Permafrost*, Edmonton, Canada, 127–131.

Onsager, L. and Runnels, L. K. (1969). Diffusion and relaxation phenomena in ice. *The Journal of Chemical Physics*, **50**(3), 1089–1103.

Pandit, B. I. and King, M. S. (1978). Influence of pore fluid salinity on seismic and electrical properties of rocks at permafrost temperatures. *Proceedings of the 3rd International Conference on Permafrost*, Edmonton, Canada, 553–566.

Paterson, N. R. and Rönkä, R. K. (1971). Five years of surveying with the Very Low Frequency Electromagnetics method. *Geoexploration*, **9**, 7–26.

Pedersen, L. B. and Becken, M. (2005). Equivalent images derived from very-low-frequency (VLF) profile data. *Geophysics*, **70**(3), 43–50.

Pedersen, L. B., Bastani, M. and Dynesius, L. (2005). Groundwater exploration using combined controlled-source and radiomagnetotelluric techniques. *Geophysics*, **70**, 8–15.

Pellerin, L. and Alumbaugh, D. (1997). Tools for electromagnetic investigation of the shallow subsurface. *The Leading Edge*, **16**, 1631–1638.

Persson, L. and Pedersen, L. B. (2002). The importance of displacement currents in RMT measurements in high resistivity environments. *Journal of Applied Geophysics*, **51**(1), 11–20.

Pfaffling, A., Haas, C. and Reid, J. E. (2004). Empirical inversion of HEM data for sea ice thickness mapping. In *10th European Meeting of Environmental and Engineering Geophysics (EAGE's Near Surface 2004)*, Extended Abstracts, Utrecht, A037.

Rozenberg, G., Henderson, J. D. and Sartorelli, A. N. (1985). Some aspects of transient electromagnetic soundings for permafrost delineation. In *CRREL Special Report*, **85–5**, 74–90.

Sartorelli, A. N. and French, R. B. (1982). Electromagnetic induction methods for mapping permafrost along northern pipeline corridors. *Proceedings of the 4th Canadian Permafrost Conference*, Canada, 283–298.

Schmöller, R. and Frühwirth, R. (1996). Komplexgeophysikalische Untersuchungen auf dem Dösener Blockgletscher (Hohe Tauern, Österreich). *Beiträge zur Permafrostforschung in Österreich. Arbeiten aus dem Institut für Geographie der Karl-Franzens-Universität Graz*, **33**, 165–190.

Scott, W., Sellmann, P. and Hunter, J. (1990). Geophysics in the study of permafrost. In *Geotechnical and Environmental Geophysics*, ed. Ward, S., Society of Exploration Geophysics, Tulsa, pp. 355–384.

Seguin, M. K. (1978). Temperature-electrical resistivity relationship in continuous permafrost at Purtuniq, Ungava Peninsula. *Proceedings of the 3rd International Conference on Permafrost*, Edmonton, Canada, 137–144.

Spies, B. R. and Frischknecht, F. C. (1991). Electromagnetic sounding. In *Electromagnetic Methods in Applied Geophysics*, ed. Nabighian, M. N., Vol. 2, Society of Exploration Geophysics, Tulsa, pp. 5–45.

Tezkan, B. (1999). A review of environmental applications of quasi-stationary electromagnetic techniques. *Surveys in Geophysics*, **20**, 279–308.

Todd, B. J. and Dallimore, S. R. (1998). Electromagnetic and geological transect across permafrost terrain, Mackenzie River delta, Canada. *Geophysics* **63**(6), 1914–1924.

Yoshikawa, K., Leuschen, C., Ikeda, A., Harada, K., Gogineni, P., Hoekstra, P., Hinzman, L., Sawada, Y. and Matsuoka, N. (2006). Comparison of geophysical investigations for detection of massive ground ice (pingo ice). *Journal of Geophysical Research*, **111**, E6, CiteID E06S19.

Zacher, G., Tezkan, G., Neubauer, F. M., Hördt, A. and Müller, I. (1996). Radiomagnetotellurics: a powerful tool for waste-site exploration. *European Journal of Environmental and Engineering Geophysics*, **1**, 139–159.

3

Refraction seismics

L. Schrott and T. Hoffmann

3.1 Introduction

The use of shallow seismic refraction in periglacial studies in the Alps dates back
to the early 1970s (Barsch 1973). At that time one-channel seismographs with a
limited number of source–receiver combinations were generally used and interpretations were limited to simple subsurface models assuming layers of a more or
less homogeneous and horizontal structure. The physical properties of frozen
ground or ice and their contrast with the unfrozen sediment (up to an order of
magnitude) stimulated many researchers to apply seismic refraction. In the 1980s,
permafrost investigations – especially the determination of active layer thicknesses – were based on simple one-channel seismic refraction using a sledgehammer as an impact source (e.g. Haeberli 1985, Barsch and King 1989, Van
Tantenhoven and Dikau 1990, King *et al.* 1992). These geophysical surveys
provided useful information regarding permafrost occurrence, and in particular
about the active layer thickness. With respect to seismic refraction, the interpretation was still usually restricted to the intercept method (see below). In many
cases, complex or more heterogeneous subsurface structures were not adequately
modelled, because of less powerful computer facilities and time-consuming
calculations.

In the 1990s, the use of multi-channel seismographs and new interpretation
software led to the application of sophisticated interpretation tools including the
Hagedoorn plus-minus method, the generalised reciprocal method (GRM), the
wavefront inversion method (WFI), network-raytracing and refraction tomography (Palmer 1981, Vonder Mühll 1993, Knödel *et al.* 1997, Reynolds 1997).
These methods allow the interpretation of structured refractors and layers with
varying P-wave velocities.

Applied Geophysics in Periglacial Environments, eds. C. Hauck and C. Kneisel. Published by Cambridge
University Press. © Cambridge University Press 2008.

Over the past decade a variety of sophisticated interpretations and visualisations of the subsurface have been applied in several geomorphic studies (Vonder Mühll 1993, Hecht 2000, Hauck 2001, Hoffmann and Schrott 2002). Besides permafrost prospecting, wide spectrums of geological and geomorphological settings and complex subsurface structures have been investigated. Applications of seismic refraction have been reported from karst and loess covered landforms, valley fill deposits (talus slopes, alluvial fans and plains, avalanche cones etc.), block fields and landslides (Hecht 2000, Tavkhelidse *et al.* 2000, Hoffmann and Schrott 2002, Schrott *et al.* 2003, Bichler *et al.* 2004).

Comparisons between different interpretation tools or different geophysical methods applied at the same geomorphological location are not as common but have been increasingly applied in recent years (King *et al.* 1988, Schrott *et al.* 2000, Kneisel and Hauck 2003, Hauck *et al.* 2004, Otto and Sass 2005, Schrott and Sass 2008, Maurer and Hauck 2007; see also Chapters 9 and 10).

Seismic refraction surveys can provide substantial information in differentiating sediment layers or for calculating sediment volumes. Besides various problems connected to the acquisition of the traveltime data (see Section 3.4), a major difficulty is the choice of the appropriate interpretation method to process the raw data (see Section 3.5).

After an introduction to measurement principles, this chapter focuses on the principles and problems of seismic refraction and gives guidelines to correct data acquisition and data processing. Finally, potential applications in periglacial and glacial terrain are briefly mentioned. Further examples are presented in subsequent chapters and in literature reviews (Scott *et al.* 1990, Vonder Mühll 1993, Hauck 2001, Vonder Mühll *et al.* 2001, Isaksen *et al.* 2002, Vonder Mühll *et al.* 2002, Hauck and Vonder Mühll 2003, Kneisel and Hauck 2003).

3.2 Measurement principles

3.2.1 Elastic waves

Investigation of the subsurface using seismic methods is based on compression waves that transmit energy by the vibration of rock and soil. Seismic waves are the result of a *stress* ε applied by an external force F across an area A to the ground surface. The vibration of the particles leads to a temporal deformation of the subsurface. Therefore the stressed subsurface material undergoes *strain*, which is the amount of deformation expressed as change in length per unit length. Away from the immediate vicinity of the seismic source the vibrations of the rock and soil can be regarded as *elastic waves*, leaving the ground unchanged after their passage. According to Hooke's law the strain caused by elastic waves is proportional to the applied stress, where the proportionality constant, called the *elastic modulus*, is a specific property of the rocks and soils within the ground.

There are two types of seismic waves that travel through the ground. While *longitudinal* or *primary (P-) waves* are characterised by deformation parallel to direction of wave propagation, the particle motion of *transverse* or *secondary (S-) waves* takes place at a right angle to the direction of wave propagation. Based on the different types of deformations the proportionality constant between strain and stress varies. In the case of P-waves the elongational modulus *j* is appropriate and for S-waves the shear modulus μ.

3.2.2 Seismic velocities

The seismic velocity is the rate at which seismic waves propagate through rocks and soils. It is dependent on the elastic modulus and the density ρ of the material through which the seismic waves propagate:

$$V_P = \sqrt{(j/\rho)} \quad \text{and} \quad V_S = \sqrt{(\mu/\rho)}, \tag{3.1}$$

where V_P and V_S are the velocities of the P- and S-waves, respectively. Because the elongational modulus $j = K + \frac{4}{3}\mu$ (with the bulk modulus $K > 0$) is always greater than the shear modulus μ, P-waves travel faster through the same medium than associated S-waves. While in solid rocks (small K) $V_P/V_S \approx 1.3$, the P-wave velocity in soils and sediments (large K) may be 3–12 times higher than the S-wave velocity. Consequently, in seismic refraction studies that are based on the detection of the first arrivals of the seismic waves, the velocities of the P-waves are estimated.

The equations for V_P and V_S suggest that velocities decrease with increasing density. The reverse, however, is true, because of decreasing elasticity of the materials with increasing density.

Regarding the application of seismic surveys in periglacial or high mountain areas, special attention should be drawn to the velocities in materials that are common in these environments and landforms (Figure 3.1). The large range of observed velocity values, spanning from ~300 m/s (loose debris) up to 6500 m/s (compact rocks) is favourable to the application of refraction seismics, since large velocity contrasts between the underlying materials are necessary. However ranges of P-wave velocities for identical or similar rocks and sediments can overlap significantly. For instance, seismic velocities for dolomite and limestone can both vary between 2000 and 6500 m/s, depending on the grade of fracturing and weathering. Consequently, there is no unambiguous relationship between certain materials and P-wave velocities and therefore it is not always possible to estimate subsurface materials simply based on seismic velocities. To differentiate, for instance, glacial till (without ice), frozen ground or solid rock,

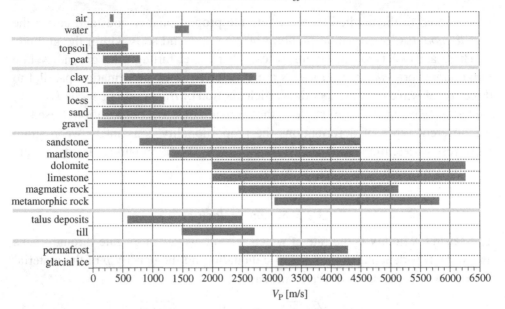

Figure 3.1.   Ranges of P-wave velocities of rocks and sediments common in high mountainous terrain (modified after Hecht 2000).

cross-checks with other geophysical methods and/or evidence from borehole logging are needed. For the same reason, the application of seismic surveys on alpine sediment storage types may lead to similar seismic velocities. The differentiation of sediment deposits that are built up of similar sediments but deposited by different geomorphic processes must fail (Hoffmann and Schrott 2003).

3.2.3 Raypath geometry

The propagation of seismic waves through layered ground is determined by the *reflection* and *refraction* of the waves at the interface between different layers. When the wave reaches the interface some energy of the wave is refracted into the deeper layer, while the reflected wave transmits the energy back into the overburden layer. Regarding the reflected wave, the angle of incidence is equal to the angle of reflection, independent of the seismic velocities of the layers. However, the angle of refraction follows *Snell's law*, which is

$$\sin \theta_i / \sin \theta_r = V_1/V_2, \tag{3.2}$$

where θ_i is the angle of incidence, θ_r is the angle of refraction and V_1 and V_2 are the seismic velocities of the upper and lower layers respectively. In the case of *critical refraction* $\theta_i = \theta_c$ and $\theta_r = 90°$, the refracted wave travels parallel to the interface with a velocity V_2 (Figure 3.2a). The critical refraction is given by the

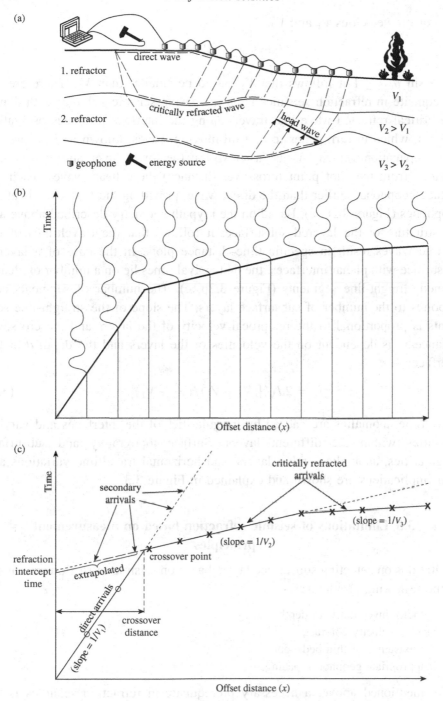

Figure 3.2. (a) Design of a seismic survey and simplified raypath geometry in a layered subsurface with three layers. (b) Corresponding seismogram resulting from a forward shot close to the first geophone. (c) Time–distance plot of traveltimes of the first arrivals extracted from the seismogram above.

ratio of the velocities V_1 and V_2:

$$\sin \theta_{c1,2} = V_1/V_2. \tag{3.3}$$

Since $\sin \theta_{c1,2} \leq 1$ it follows that V_1 has to be smaller than V_2. Therefore the prerequisite in refraction seismics is an increasing seismic velocity with depth. The critical refracted wave that travels along the interface produces oscillation stresses, which in turn generate upward-moving waves, known as *head waves*. The propagation velocity V_2 is faster within the lower layer and at a certain distance from the shot point (crossover distance) these head waves reach the surface geophones earlier than the direct wave, providing the first arrivals at the geophones (Figure 3.2b, c). Based on the raypath geometry described above and the structure of the layered subsurface, it follows that the traveltimes of the seismic waves result in specific time–distance plots. In the case of a layered subsurface with planar interfaces, the first arrival times lie on a number of clearly defined straight-line segments (Figure 3.2b, c). The number of segments corresponds to the number of subsurface layers. The slope of the straight-line segments is proportional to the reciprocal velocity of the layers and the crossover distance x_c is dependent on the velocities of the layers and the depth d of the interface:

$$x_c = 2d\sqrt{[(V_2 + V_1)/(V_2 - V_1)]}. \tag{3.4}$$

Traveltime anomalies are caused by irregularities of the interfaces and varying velocities within the different layers. Surface topography and subsurface irregularities, anomalies within layers, and horizontal traveltime variations and their implications are shown and explained in Figure 3.3.

3.3 Limitations of seismic refraction based on measurement principles

Limitations on detecting subsurface layers based on seismic refraction are caused by the following problems:

(i)   velocity inversion with depth,
(ii) lack of velocity contrast,
(iii) the existence of thin beds, and
(iv) inappropriate geophone spacing.

As mentioned above, a necessary prerequisite in refraction seismics is the increasing seismic velocity with depth. In the case of low-velocity layers the waves are refracted into the ground and no head waves are produced. Examples of velocity inversion with depth in mountainous areas may be peat layers overlain

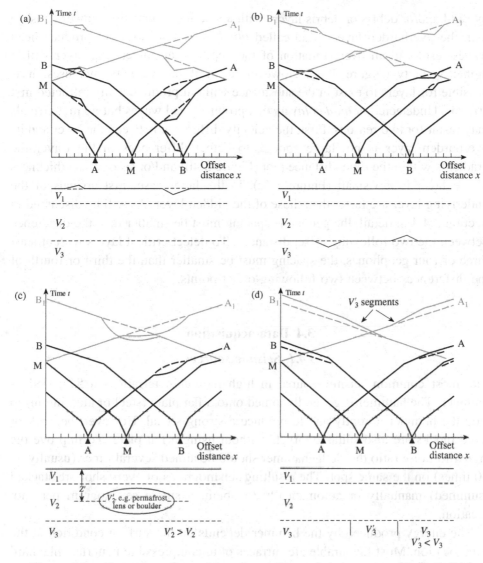

Figure 3.3. Examples of traveltime anomalies and their causes. Note: Solid lines indicate traveltimes with anomalies and dashed lines indicate regular traveltimes. The letters A, B and M indicate the position and corresponding traveltimes of the forward, reverse and medium shots, respectively. Letters A_1 and B_1 indicate traveltimes of forward and reverse shots outside the spread. (a) Irregularities of surface topography influence traveltime through all straight-line segments (velocities) at identical offset distances. (b) Irregularities of subsurface topography of the third layer (V_3) are shown in traveltime anomalies of the corresponding refractor only. (c) Anomaly within the subsurface layer, e.g. permafrost lens or boulder, with higher velocity (V'_2) than the surrounding layer (V_2). Only traveltimes of layers below the anomaly are affected. (d) Horizontal variation of refracted velocities (e.g. bedrock–sediment–bedrock). Only traveltimes of the corresponding layer are affected resulting in larger traveltimes (modified after Reynolds 1997).

by sand and/or debris or debris layers with a smaller compaction (lower density) than the overburden layer. A so called *blind layer* that does not produce head waves results in an overestimation of the depth of an underlying layer with a higher velocity (Figure 3.4a). However, even if the velocity increases, it is possible for layers to exist in the subsurface without producing any refracted first arrivals. Undetected or *hidden layers* that produce head waves but no first arrivals may result for two reasons. First, the velocity difference between the layer and its overburden layer is not large enough to detect differences in the traveltime gradients within the time–distance plot (Figure 3.4b) and/or second, the thickness of the layer is too small (Figure 3.4c). In the latter case, first arrivals of the underlying layer are faster than those of the hidden layer. As will be discussed in Section 3.4.4 in detail, the geophone spacing must be smaller than the difference between the two following offset distances. To detect critical layers with at least three or four geophones, the spacing must be smaller than the third or fourth of the differences between two following offset points.

3.4 Data acquisition

3.4.1 Seismic sources

The most common seismic source in high mountain terrain is a 5 kg sledge-hammer. The hammer is generally aimed onto a flat plate (steel or aluminium) to stop the hammer abruptly and to produce a strong signal. However, bedrock or fixed blocks are also suitable targets without using a flat plate. To improve the signal-to-noise ratio the sledgehammer should be aimed several times (usually 5–10 times) on the same spot. The resulting seismograms of every shot are stacked (summed) manually or automatically to obtain a single seismogram per shot location.

The energy produced by the hammer depends on the surface condition at the shot location. Most favourable are surfaces of uncompressible materials like hard rock, compacted sediments, etc. On highly compressible surface materials (e.g. peat or other organic rich materials) or unconsolidated coarse debris the energy propagation into the ground is limited.

Generally, a sledgehammer is sufficient to measure the first arrivals along offset distances of approximately 50 m and seldom more than 90 m. This corresponds to penetration depths of 10 to 30 m. More powerful sources must be used for longer surveys with larger penetration depths. However, the application of more powerful seismic sources in mountain terrain is limited by the higher weight of the equipment. Therefore, on steep terrain the use of drop weights of up to 100 kg (as used in flat-terrain seismic refraction) is not possible.

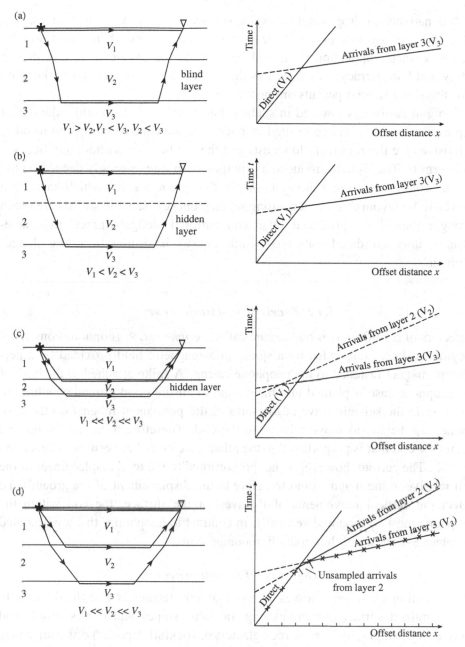

Figure 3.4. Synthetic three-layer models (left) and corresponding traveltimes (right). Examples showing limit of seismic refraction. (a) Velocity inversion of the second layer that is not detected in traveltimes. (b) Velocity contrast between layers one and two is too small to be detected in traveltimes. (c) Undetected second layer due to insufficient thickness. (d) Undetected second layer due to insufficient geophone spacing. Geophone spacing is indicated by the ticks on the *x*-axis of the traveltime diagram (modified after Reynolds 1997).

Alternatively to drop weights, explosives are common seismic sources. The advantages of explosives are their immense power and their small weight. However, their application is limited because of the problems arising due to safety and bureaucracy. In most cases the work must be supervised by licensed shot-firers and special permits are necessary.

Shotgun cartridges lowered in shallow boreholes (0.5–1 m) allow the use of explosives in a safe and controlled manner. The advantages compared to other explosives are the relatively low costs and the application without any licenses and permits. The signals produced by shotgun cartridges strongly depend on the ground conditions around the borehole. In fine-grained soils, which have been proven to be favourable for generating seismic energy, seismic signals are much stronger than those produced by a conventional sledgehammer. In coarse-grained, unconsolidated soils the seismic energy is similar to a 5 kg sledge-hammer.

3.4.2 Receiver of seismic waves

Detectors of seismic waves on land are called *geophones*. A geophone consists of a cylindrical coil suspended by a spring in a magnetic field provided by a permanent magnet fastened to the geophone casing. A spike attached to the base of the geophone case is planted into the ground to ensure good ground coupling. In response to the seismic wave at the surface, the permanent magnet on the geophone case begins to move relative to the coil. Therefore a current is induced within the coil that is proportional to the relative velocity between the magnet and the coil. The current, however, is not proportional to the total displacement of the coil relative to the magnet (and therefore to the displacement of the ground). To detect the vertical movements of P-waves, at the surface, the coil within the geophone must be mounted vertically, in contrast to geophones that are designed to detect S-waves with horizontally mounted coils.

Detection of seismic waves

The detection of seismic waves in mountainous terrain is a difficult task. In coarse-grained surface conditions (e.g. on talus slopes and debris cones) and hummocky topography (e.g. on rock glaciers or rockfall deposits) difficulties may arise from weak coupling of geophones to the subsurface. To detect seismic waves, for instance on talus slopes with larger block sizes directly on the talus surface, the coupling of the geophones is improved when the upper layer of larger boulders is removed and the spikes of the geophones are pushed into the fine-grained material underneath. To couple geophones on large boulders or on compact bedrock it is often recommended to drill into the rock and to plunge the spike within the drill hole. Strong coupling of the geophones, however, is crucial in order to receive clear signals with high signal-to-noise ratios.

In the neighbourhood of torrents, in strong wind or rainfall the detection of seismic waves may be impossible due to the strong noise in the seismic record.

3.4.3 Seismographs: recorders of seismic waves

Seismographs record the seismic signals detected by the geophones. However, there is a wide variety of seismographs available ranging from simple one-channel seismographs, with a basic graphical display, to complex units with 48 channels (or more). Sophisticated units are able to digitise, filter and stack signals from single shots and to store signals for different numbers of detectors. New instruments are typically entirely menu-driven and controlled via an internal microprocessor with preloaded software for data acquisition and simple data processing or an external laptop operating under Windows 2000/XP. Distributed systems allow several seismograph units to be combined, thereby increasing the number of geophones for seismic reflection surveys or 3D refraction surveys.

To apply seismic surveys in high mountain terrain it is highly recommended to use seismographs with several single light-weight units rather than to use an all-in-one (heavy) seismograph.

3.4.4 General considerations for shot–receiver geometries

For a seismic refraction survey the geophones are laid out along a cable that follows a straight line. This set of geophones constitutes the so called *spread* (Figure 3.2). For practical reasons geophones are equally spaced along the spread. The distance between the shot point and a geophone is known as the offset distance. The longest offset distance and the geophone spacing determine the depth of penetration and resolution of the seismic record for a given subsurface. Assuming a sufficiently strong seismic source, large offset distances generally result in large penetration depths. As a simple rule of thumb the depth of penetration is about 1/3–1/5 of the offset distance. Taking the limited power of a sledgehammer into account it is possible to measure the first arrivals of the seismic waves up to 90 m from the seismic source. By rule of thumb, it follows that seismic surveys are able to estimate the structure of the subsurface within the upper 30 m. For example, using 24 geophones (with a 24 channel seismograph) combined with a spacing of 4 m results in a spread of 92 m. If the expected depths of the interfaces are shallower than 30 m the geophone spacing may be reduced (i.e. 1–3 m) to gain a higher resolution record of the subsurface. Reduced geophone spacing improves the estimate of the crossover distance. To detect a layer within the time–distance plot the geophone spacing must be smaller than the difference between two following crossover distances. In cases of thin subsurface layers, where the distance between crossover points is small, a narrow geophone spacing (1–3 m) is necessary (Figure 3.4).

In addition to the geophone spacing and survey distance it is important to consider the number of shot points per survey. In the case of a layered subsurface with horizontal planar interfaces, a single shot point can often be sufficient. A dipping layer, however, makes it necessary to carry out forward and reverse shooting. This enables the determination of all parameters required for refraction geometry modelling (see Section 3.5). Under complex subsurface conditions, a large number of shot points (in the case of 24 channels, 10 to 20 shots are appropriate) may be helpful for a detailed analysis of the subsurface structure (see Section 3.5.2). For all first arrivals to have come via the refractors it is necessary to place shot points between the geophones. This additional effort is easy to carry out and it is strongly recommended always to cover a survey line with several shot points (i.e. 10–20). In particular, the inversion of traveltimes to subsurface velocity models using 2D refraction tomography algorithms requires as many source points as possible (Hauck 2001).

3.4.5 Annotation of field records

The annotation of the field records is an important part of the fieldwork in order to process and interpret the data and to prevent any misinterpretation of the data.

Annotations should include general information about the date, observer, spread number and weather conditions during the field record and previous to the fieldwork. Additionally the location of the spread (especially the first and last geophones) and its topography as well as its geomorphological context must be described. The location of the spread may be mapped on a topographic map or be estimated using a GPS. To measure the topography along a spread accurately, a ruler in combination with slope measuring tools or a theodolite should be used. The precise description of the geomorphological context of the spread significantly improves the interpretation of subsurface models.

Considering data acquisition, information about geophone spacing, number of shots combined in a single record, shot number and locations are necessary for data processing.

In addition to the annotation of field records, high-quality hard copies of the seismograms should always be taken either in the field or as soon as possible afterwards. Hard copies may allow the subsequent processing and interpretation of surveys if data are lost.

3.5 Data processing

In this section different procedures for data processing are described. The processing follows the general steps:

(i) extraction and interpretation of traveltimes from measured seismograms, and
(ii) inversion of the traveltimes to P-wave velocity models of the subsurface.

We describe the extraction and interpretation of traveltimes as a prerequisite to the inversion of the traveltimes. Furthermore the basic principles of the selected inversion methods are explained. Besides the traditional intercept-time method we focus on more sophisticated techniques like the wavefront inversion method, network raytracing and 2D-refraction tomography. Detailed descriptions of the inversion methods can be found in Kearey *et al.* (2002), Knödel *et al.* (1997), Reynolds (1997).

3.5.1 Extraction of traveltimes

To start with data processing, general procedures must be applied independently of the inversion method. Since all inversion methods are based on traveltimes of the first arrivals, the following steps are a prerequisite to the inversion process:

(i)   importing the seismic data into processing software;
(ii) picking the first arrivals;
(iii) creating time–distance plots.

The import of seismic data into the processing software strongly depends on the particular software used (see your software handbook for details).

In the case of high signal-to-noise ratios, picking the first arrivals on refraction records is a straightforward task (Figure 3.5a). However, if the signal-to-noise

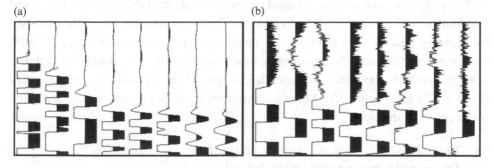

(a)   (b)

Figure 3.5. Parts of multi-channel refraction seismograms. (a) Example with clear first arrivals and high signal-to-noise ratio. (b) Example of weak first arrivals and low signal-to-noise ratio.

ratio is poor, especially at remote geophones, picking the first arrivals is prone to
significant errors. Comparing weak first arrivals with the same wavefront of
traces with stronger signals in the near neighbourhood may help to gain a
complete set of first arrivals even for seismograms where the signal-to-noise ratio
is poor (Figure 3.5b). Generally, the traveltimes of traces with larger offset
distances should be longer than those with smaller offset distances. Exceptions to
this rule may occur if the geophones are not in a straight line due to irregularities
in the surface topography (Figure 3.3a) or in the case of subsurface anomalies
(Figure 3.3b). However, if exceptions occur they should be consistent in every
shot taken along the spread.

Plotting traveltimes against distance along the spread allows one to check the
consistency of traveltimes, to evaluate the influence of topography, and to
interpret the subsurface structure (Figure 3.6; see also Chapter 9). In the case of a
horizontally structured subsurface the time–distance plots of every shot must be
almost identical. Even though nature is not horizontally structured in homogeneous layers, the traveltimes of neighbouring shots should follow the same trend.
Another important criterion of the quality of the picked first arrivals is the total
traveltime of the forward (shot taken at point A and geophone located at B) and
reverse shot (shot taken at B and geophone located at A). As the raypaths of the
forward and reverse shots are the same, it follows that the total traveltime must be
the same.

3.5.2 Traveltime inversion

A large number of methods exist to invert traveltimes into P-wave velocity
models of the subsurface. They range from the simple/restrictive intercept-time
method to the complex refraction tomography that gives a detailed picture of
subsurface anomalies.

The *intercept-time method* is based on the estimation of the intercept times and
apparent velocities of the refracted arrivals. These values are generally obtained
by drawing best-fit lines through the linear segments of the time–distance plot.
The intercept time and the apparent velocity are defined as the time at which the
best-fit line cuts the time axis and the reciprocal of its gradient, respectively
(Figure 3.2c and Figure 3.7). Some assumptions must be met in order to apply the
intercept time method:

(i)   The subsurface consists of n layers with constant P-wave velocities $V_1, \ldots ,V_n$.
(ii) The velocities increase with depth, that is $V_{n+1} > V_n$.
(iii) The interfaces between the layers are planar and their gradient relative to the surface
      is smaller than $10°$.

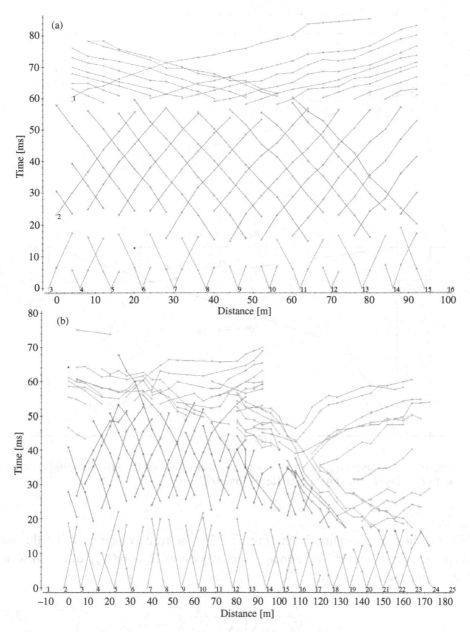

Figure 3.6. Combined time–distance plot and assigned traveltimes. Traveltimes with the same tone correspond to the same layer. (a) Time–distance plot of a survey with minor surface and refractor topography. The corresponding refraction sounding RS40 is located on an almost flat alluvial plain within the Reintal (Bavarian Alps, Germany) (from Hoffmann and Schrott 2003). (b) Composite time–distance plots (two 90 m spreads) with indications of a major subsurface anomaly. The traveltimes were measured on a rectilinear slope in the Turtmann Valley, Swiss Alps (from Otto and Sass 2006).

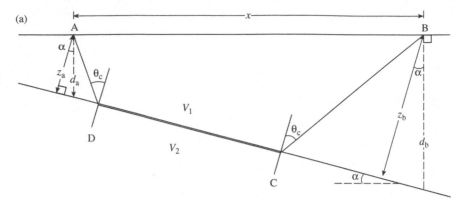

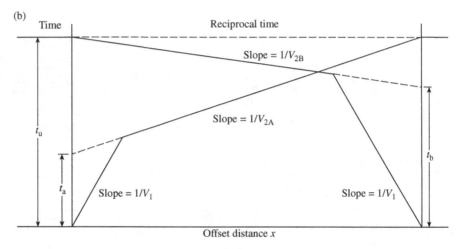

Figure 3.7.   Raypath geometry (a) and traveltimes (b) of the layer-dipping case with two layers (modified after Reynolds 1997).

In the case of a dipping layer (Figure 3.7), the true velocities V_n of the nth layers are approximated based on apparent velocities V_{nA} and V_{nB} of the forward shot A and reverse shot B:

$$\frac{1}{V_n} \approx \frac{1}{2}\left(\frac{1}{V_{nA}} + \frac{1}{V_{nB}}\right). \tag{3.5}$$

The depth Z of the interface n under the shot point A is calculated by

$$Z_{n-1,A} = \left(t_{n-1,A} - \sum_{j=1}^{n-2}\frac{2Z_{jA}}{V_{jn}}\right)\frac{v_{n-1}}{2\cos(\theta_{cn-1,n})}, \tag{3.6}$$

where $\theta_{cn-1,n}$ is the critical fraction angle calculated by Equation (3.1) and V_{jn} is the depth inversion factor:

$$V_{jn} = \frac{V_j V_n}{\sqrt{V_n^2 - V_j^2}}. \tag{3.7}$$

The depth under shot point B is calculated analogously. The resulting model of the subsurface is a planar model with homogeneous layers. Surface and refractor topography are not considered in the intercept time analysis. However, if the subsurface shows varying velocities within the layers and a nonlinear topography of the refractors, the intercept-time method gives mean layer thicknesses and mean seismic velocities. Because of its rapidity the intercept-time method is useful for a first field interpretation of the seismic surveys and to develop a more detailed survey plan in the field. For homogeneous and quasi-horizontally layered subsurface structures the assumptions of the intercept-time method are met. In difficult and more complex terrain, the results may vary significantly, sometimes exceeding 50%, and as a consequence wrong geomorphological conclusions can be drawn from too simplistic geophysical interpretations. In such a case, simple interpretations (e.g. using the intercept-time method) remain unsatisfactory and will not provide a reliable model. Schrott *et al.* (2000) compared refractor depths from a blockfield using the intercept method and GRM, resulting in differing results of the intercept method compared to the more sophisticated GRM due to the uneven surface of the refractors. The authors suggest that modelling results obtained using the intercept-time method have to be treated carefully and emphasise the necessity of complementary and highly sophisticated interpretation tools to get more reliable models.

To model complex subsurface structures, more sophisticated methods are necessary. The *wavefront inversion method* (WFI) allows the inversion of traveltimes in the case of considerable surface and refractor topography. The inversion method is based on a finite difference approximation of the eikonal equation that migrates the combined forward and reverse traveltimes into depth (e.g. within the commercially available software program Reflex-W, Sandmeier 2002).

In order to produce subsurface models based on the WFI the following steps are necessary (Sandmeier 2002):

(i)   Extraction of the traveltimes of the first arrivals of the direct and refracted P-waves.
(ii) Combining traveltimes of each record to a single time–distance plot.
(iii) Interpretation of the traveltimes and determination of the number of layers of the subsurface model. To improve the interpretation of the traveltimes, anomalies of the surface topography should be considered (Figure 3.3a).
(iv) Assignment of the traveltimes to the corresponding layers of the subsurface model (different tones in Figure 3.6).

(v)   Inversion of the traveltimes of the first layer resulting in a P-wave velocity model of the first layer. Surface irregularities can be included into the P-wave velocity model at this stage.

(vi)   Extraction of forward and reverse traveltime segments for the next layer based on chosen forward and reverse shot.

(vii) Inversion of the traveltime segments based on the wavefront inversion algorithm and the overburden model.

Steps (vi) and (vii) are repeated until all assigned traveltimes are inverted into a P-wave velocity model with corresponding numbers of layers. Therefore the WFI is an iterative process, meaning that each layer must be inverted separately and based on the existent overburden model. Thus the (re)construction of the subsurface model begins with the uppermost layer and ends up with the basal layer.

As mentioned above, the WFI method allows inverting traveltimes to subsurface models of any complexity. The inversion, however, is limited because only a small part of the traveltime data is used and because the first arrivals must be assigned to the subsurface layers with sufficient accuracy. The assignment of traveltimes to subsurface layers under complex conditions is not always straightforward. It is therefore highly recommended to evaluate the WFI models using a *network raytracing* algorithm (e.g. within Reflex-W; Sandmeier 2002) or similar inversion programs. Network raytracing calculates synthetic traveltimes of a given model based on the raypaths with the shortest traveltimes from a defined shot point along a given spread. To evaluate WFI models using network raytracing, the following steps are performed (Hoffmann and Schrott 2003):

(i)   Choosing an initial model (typically an existing WFI model).

(ii)   Loading the measured traveltimes.

(iii) Calculation of synthetic traveltimes.

(iv)   Visual comparison of synthetic and measured traveltimes and manual modification of the subsurface model, by changing the layer depths and the velocities of the start model.

Steps (iii) and (iv) are repeated until the differences between synthetic and measured traveltimes are minimised. An application of a generated subsurface model using the WFI and network raytracing is shown in Figure 3.8 (Plate 3).

It should be kept in mind that the assignment of synthetic traveltimes to measured data is a time-consuming process and requires experience. Meanwhile *2D refraction tomography* inversion schemes have been developed to automatically adapt synthetic traveltimes to measured data (analogous to electrical resistivity tomography, see Chapter 1). The tomography inversion results in a grid model showing the two-dimensional P-wave velocity distribution of the

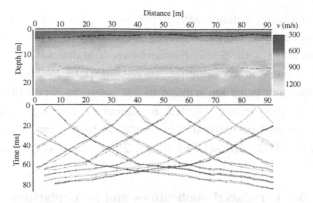

Figure 3.8. Model results of a network raytracing and tomography analysis based on seismic refraction data from an alluvial plain. Top panel: Comparison of the network-raytracing model (solid black line) and refraction tomography (colours indicate the changing velocities of the tomography model). Bottom panel: Comparison of the measured traveltimes (black lines) and calculated traveltimes (coloured crosses) (modified after Hoffmann and Schrott 2003). For colour version see Plate 3.

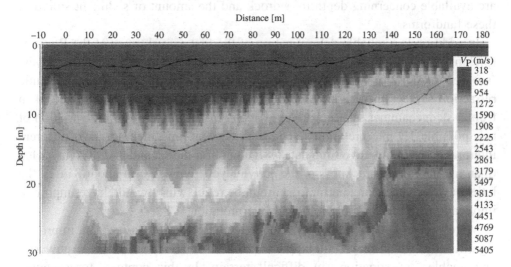

Figure 3.9. Tomography model of traveltimes measured on a rectilinear slope in the Turtmann Valley, Swiss Alps (traveltimes are shown in Figure 3.6b). For colour version see Plate 4.

subsurface (Figure 3.9 (Plate 4)). Based on an iterative adaptation (e.g. SIRT, Simultaneous Iterative Reconstruction Technique), an initially chosen velocity model is automatically adjusted in order to minimise the traveltime residuals between measured and calculated data (e.g. Lanz *et al.* 1998, Sandmeier 2002). To model continuous velocity changes – as observed in sediment with continuously increasing density with depth – the tomography inversion allows curved ray propagation. Consequently, the velocity of initial models should continuously increase with depth. Furthermore, a major advantage of the tomography is the inversion of complex traveltime data to complex subsurface structures without any prior knowledge of the subsurface conditions. Therefore refraction tomography is well suited to detecting heterogeneous subsurface structures (e.g. isolated permafrost occurrence) in high mountain terrain (Hauck 2001, Musil et al. 2002, Kneisel and Hauck 2003; see also Chapter 10).

3.6 Periglacial applications and particularities

Over the past three decades, probably the most common application of seismic refraction in periglacial terrain is the detection of permafrost (Timur 1968, Barsch 1973, Zimmerman and King 1986, King et al. 1992, Vonder Mühll 1993, Hauck 2001, Musil et al. 2002, Hauck and Vonder Mühll 2003). Other applications focus on some of the dominant geomorphic forms in alpine environments, e.g. large talus slopes, alluvial plains and alluvial fans, where few reliable data are available concerning depth to bedrock and the amount of sediment stored in these landforms.

In addition to a reliable determination of the first arrival times of the P-waves (picking), the correct assignment of the traveltime data to the respective layers (see Section 3.5.2) is probably the most difficult and error-prone step in seismic refraction surveys in periglacial terrain. On the basis of a single data set taken from a sledgehammer seismic survey, King et al. (1992) showed the variation of two possible 1D interpretations using the intercept-time method. Two different interpretations of the traveltime data assignment for the second and third layer led to an increase in the thickness of the active layer from 4.1 m to about 8 m. In the latter case the underlying permafrost might be relict, which would have important implications to the geomorphological interpretation.

More recently, data gathering and data interpretation have been improved significantly, which makes the application of seismic refraction more powerful and enables investigations in difficult terrain. In this context, tomographic inversion as a sophisticated modelling tool to investigate complex subsurface structures has become more popular in geomorphic research (Figure 3.9). Even isolated permafrost lenses or boulders can be detected and visualized using appropriate software tools (Hauck 2001, Musil et al. 2002, Kneisel and Hauck 2003, see also Chapter 10).

In contrast to periglacial environments, the application of seismic refraction on glaciers is less frequent. These studies, however, give valuable insights regarding ice properties, internal structures and sediments. Milana and Maturano (1999) used seismic refraction to measure the P-wave velocities of glacier ice in the Andes. The increase of P-wave velocity at the glacier terminus suggests an increase in glacier density due to older and less crevassed ice at the glacier terminus. The authors conclude that the lack of refracted waves, even in a 300 m long profile, indicates either a heavily weathered rock or a probable wet debris unit between glacier ice and bedrock. Benjumea *et al.* (2003) estimated the water content in a temperate glacier from seismic velocity data. Knowledge about the water content of a glacier improves the accuracy of ice-thickness measurements and has a notable influence on effective strain rates. However, water content estimates from seismic velocity data should be compared with radar data because seismic velocity is an increasing function of density, which causes higher (water content values) values, whereas radar velocity is a function of decreasing density causing lower values.

In general it is highly recommended to check the results with alternative geophysical surveys such as ground-penetrating radar or DC resistivity (see Chapters 9 and 10). Whenever possible (and available), borehole data should be included to validate the model. The successful interpretation of the seismic data is closely linked to knowledge of the geological and geomorphological setting in the area.

3.7 Checklist

Before starting a survey:

- Is the method used adequate to answer the scientific question?
- Are there any other useful complementary methods?
- What is the minimum expected number of profiles/surveys?
- Is there any information on the expected subsurface condition based on geomorphological and geological setting and interpretation?
- Is there any information about expected depth of refractors?
- Is there any possibility of hidden layers?
- Are there any other critical factors (groundwater, hollows etc.) influencing the seismic signal?

During the survey:

- Check carefully the coupling of each geophone.
- Check carefully (visually) any shot and repeat single shots if necessary.
- If a sledgehammer is used as an impact source, stack several shots (at least 3–5) to improve the signal-to-noise ratio.
- Check if the total spread and the geophone distance are appropriate in terms of resolution and/or possible depth of the lowermost refractor. Change the geophone distance if necessary.
- Make hard copies (printouts) of seismograms.

After the survey:

- Discuss your results carefully with experienced geomorphologists and geophysicists before drawing major conclusions.
- Use postprocessing software that offers different interpretation tools (e.g. Reflex).
- Make (whenever possible) cross-checks using complementary geophysical methods.
- Try to avoid the misinterpretation of critical data sets and repeat the survey if necessary.

REFERENCES

Barsch, D. (1973). Refraktionsseismische Bestimmungen der Obergrenze des gefrorenen Schuttkörpers in verschiedenen Schüttkörpern Graubündens. *Zeitschrift für Gletscherkunde und Glazialgeologie*, **9**, 143–167.

Barsch, D. and King, L. (1989). Origin and geoelectrical resistivity of rock glaciers in semi-arid subtropical mountains (Andes of Mendoza, Argentina). *Zeitschrift für Geomorphologie*, **33**, 151–163.

Benjumea, B., Macheret, Y. Y., Navarro, F. J. and Teixido, T. (2003). Estimation of water content in a temperate glacier from radar and seismic sounding data. *Annals of Glaciology*, **37**, 317–324.

Bichler, A., Bobrowsky, P., Best, M., Douma, M., Hunter, J., Calvert. T. and Burns, R. (2004). Three-dimensional mapping of a landslide using a multi-geophysical approach: the Quesnel Forks landslide. *Landslides*, **1**, 29–40.

Haeberli, W. (1985). *Creep of Mountain Permafrost: Internal Structure and Flow of Alpine Rock Glaciers*. Mitteilungen der Versuchsanstalt für Wasserbau, Hydrologie und Glaziologie, 77, 143pp.

Hauck, C. (2001). *Geophysical Methods for Detecting Permafrost in High Mountains*. Mitteilungen der Versuchsanstalt für Wasserbau, Hydrologie und Glaziologie, 171, 204pp.

Hauck, C. and Vonder Mühll, D. (2003). Evaluation of geophysical techniques for application in mountain permafrost studies. *Zeitschrift für Geomorphologie, Supplement*, **132**, 161–190.

Hauck, C., Isaksen, K., Vonder Mühll, D. and Sollid, J. L. (2004). Geophysical surveys designed to delineate the altitudinal limit of mountain permafrost: an example from Jotunheimen, Norway. *Permafrost and Periglacial Processes*, **15**(3), 191–205.

Hecht, S. (2000). Fallbeispiele zur Anwendung refraktionsseismischer Methoden bei der Erkundung des oberflächennahen Untergrundes. *Zeitschrift für Geomorphologie, Supplement*, **123**, 111–123.

Hoffmann, T. and Schrott, L. (2002). Modelling sediment thickness and rockwall retreat in an Alpine valley using 2D-seismic refraction (Reintal, Bavarian Alps). *Zeitschrift für Geomorphologie, Supplement*, **127**, 175–196.

Hoffmann, T. and Schrott, L. (2003). Determining sediment thickness of talus slopes and valley fill deposits using seismic refraction – a comparison of 2D interpretation tools. *Zeitschrift für Geomorphologie, Supplement*, **132**, 71–87.

Isaksen, K., Hauck, C., Gudevang, E., Oedegaard, R. S. and Sollid, J. L. (2002). Mountain permafrost distribution in Dovrefjell and Jotunheimen, southern Norway, based on BTS measurements and 2D tomography data. *Norsk Geografisk Tidsskrift*, **56**, 122–136.

Kearey, P., Brooks, M. and Hill, L. (2002). *An Introduction to Geophysical Exploration*. Blackwell.

King, L., Gorbunov, A. P. and Evin, M. (1992). Prospecting and mapping of mountain permafrost and associated phenomena. *Permafrost and Periglacial Processes*, **3**(2), 73–81.

King, M. S., Zimmerman, R. W. and Corwin, R. F. (1988). Seismic and electrical properties of unconsolidated permafrost. *Geophysical Prospecting*, **36**, 349–364.

Kneisel, C. and Hauck, C. (2003). Multi-method geophysical investigation of a sporadic permafrost occurrence. *Zeitschrift für Geomorphologie, Supplement*, **132**, 145–159.

Knödel, K., Krummel, H. and Lange, G. (1997). *Geophysik. Handbuch zur Erkundung des Untergrundes von Deponien und Altlasten*. Springer.

Lanz, E., Maurer, H. R., and Green, A. G. (1998). Refraction tomography over a buried waste disposal site. *Geophysics*, **63**(4), 1414–1433.

Maurer, H. and Hauck, C. (2007). Geophysical imaging of alpine rock glaciers. *Journal of Glaciology*, **53**(180), 110–120.

Milana, J.-P. and Maturano, A. (1999). Application of radio echo sounding at the arid Andes of Argentina: the Agua Negra Glacier. *Gobal and Planetary Change*, **22**, 179–191.

Musil, M., Maurer, H., Green, A. G., Horstmeyer, H., Nitsche, F., Vonder Mühll, D. and Springman, S. (2002). Shallow seismic surveying of an Alpine rock glacier. *Geophysics*, **67**(6), 1701–1710.

Otto, J. C. and Sass, O. (2006). Comparing geophysical methods for talus slope investigations in the Turtmann valley (Swiss Alps). *Geomorphology*, **76**, 257–272.

Palmer, D. (1981). An introduction to the generalized reciprocal method of seismic refraction interpretation. *Geophysics*, **46**(11), 1508–1518.

Reynolds, J. M. (1997). *An Introduction to Applied and Environmental Geophysics*. John Wiley & Sons.

Sandmeier, K. (2002). *Reflex-W, version 2.5.9*. Sandmeier Scientific Software, Karlsruhe, Germany.

Schrott, L. and Sass, O. (2008). Application of field geophysics in geomorphology: advances and limitations exemplified by case studies. *Geomorphology*, **93**, 55–73.

Schrott, L., Pfeffer, G. and Möseler, B. M. (2000). Geophysikalische Untersuchungen an einer Blockhalde im Mittelgebirge (Hundsbachtal, Eifel). *Acta Universitatis Purkynianae Ústi nad Labem, studia biologica* **4**, 19–31.

Schrott, L., Hufschmidt, G., Hankammer, M., Hoffmann, T. and Dikau, R. (2003). Spatial distribution of sediment storage types and quantification of valley fill deposits in an upper Alpine basin, Reintal, Bavarian Alps, Germany. *Geomorphology*, **55**, 45–63.

Scott, W., Sellmann, P. and Hunter, J. (1990). Geophysics in the study of permafrost. In *Geotechnical and Environmental Geophysics*, ed. Ward, S., Society of Exploration Geophysics, Tulsa, pp. 355–384.

Tavkhelidse, T., Schulte, A., Stumböck, M. and Schuhkraft, G. (2000). Aufbau und Entwicklung der Schuttkegel im Finkenbachtal, Südlicher Odenwald. *Jenaer Geographische Schriften*, **9**, 95–110.

Timur, A. (1968). Velocity of compressional waves in porous media at permafrost temperatures. *Geophysics*, **33**(4), 584–595.

Van Tantenhoven, F. and Dikau, R. (1990). Past and present permafrost distribution in the Turtmanntal, Wallis, Swiss Alps. *Arctic and Alpine Research*, **22**(3), 302–316.

Vonder Mühll, D. (1993). *Geophysikalische Untersuchungen im Permafrost des Oberengadins.* Mitteilungen der Versuchsanstalt für Wasserbau, Hydrologie und Glaziologie, 122, 222pp.

Vonder Mühll, D., Hauck, C., Gubler, H., McDonald, R. and Russill, N. (2001). New geophysical methods of investigating the nature and distribution of mountain permafrost. *Permafrost and Periglacial Processes,* **12**(1), 27–38.

Vonder Mühll, D., Hauck, C. and Gubler, H. (2002). Mapping of mountain permafrost using geophysical methods. *Progress in Physical Geography,* **26**(4), 640–657.

Zimmerman, R. W. and King, M. S. (1986). The effect of freezing on seismic velocities in unconsolidated permafrost. *Geophysics,* **51**, 1285–1290.

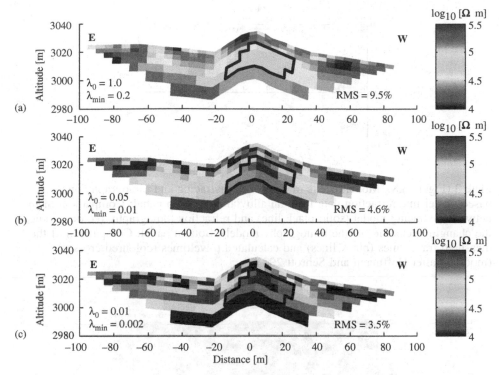

Plate 1 (Figure 1.4). Tomographic inversions of data set recorded on ice-cored moraine near Zermatt for (a) large, (b) intermediate and (c) small damping values. The resistive zone interpreted to be the ice core is marked by a solid line (from Hauck and Vonder Mühll 2003b).

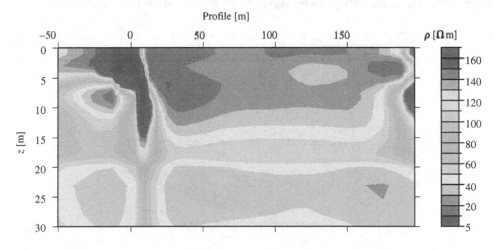

Plate 2 (Figure 2.8). Resistivity section obtained from a 2D inversion of RMT data (from Hördt *et al.* 2000).

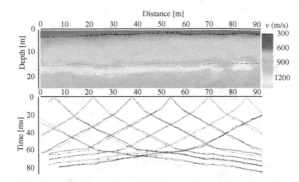

Plate 3 (Figure 3.8). Model results of a network raytracing and tomography analysis based on seismic refraction data from an alluvial plain. Top panel: Comparison of the network-raytracing model (solid black line) and refraction tomography (colours indicate the changing velocities of the tomography model). Bottom panel: Comparison of the measured traveltimes (black lines) and calculated traveltimes (coloured crosses) (modified after Hoffmann and Schrott 2003).

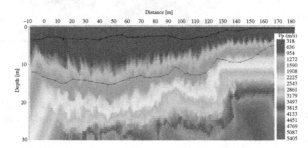

Plate 4 (Figure 3.9). Tomography model of traveltimes measured on a rectilinear slope in the Turtmann Valley, Swiss Alps (traveltimes are shown in Figure 3.6b).

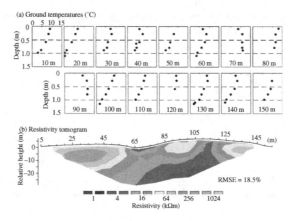

Plate 5 (Figure 6.3). (a) Profiles of shallow ground temperature (°C) and (b) inverted DC resistivity (kΩ m) along the same survey line (originally presented in Ishikawa 2003).

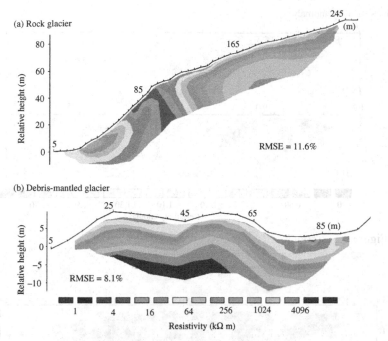

Plate 6 (Figure 6.4). Inverted DC resistivity tomogram of (a) a rock glacier and
(b) the debris-mantled part of the Kanchenjunga Glacier. The unit electrode spacing is
5 m for both measurements (originally presented in Ishikawa *et al.* 2001).

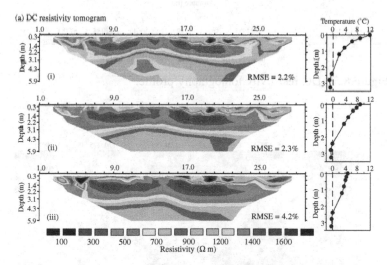

Plate 7 (Figure 6.5). Inversion results of ERT imaging at the permafrost-underlying talus
slope, northeast Mongolia. (a) DC resistivity (Ω m) tomogram on (i) 1 July, (ii) 24 August
and (iii) 20 September 2003. Daily means of ground temperatures on the respective days
are also shown. Unit electrode spacing was 1m and total number of electrodes was 30.
(b) Plots of anomaly index for two consecutive resistivity measurements, showing resistivity
changes (i) between 1 July and 24 August, and (ii) between 24 August and 20 September 2003.

(b) Resistivity anomaly

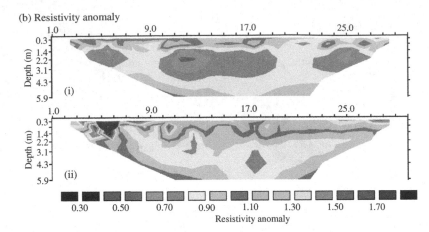

Plate 7 (Figure 6.5). (cont.)

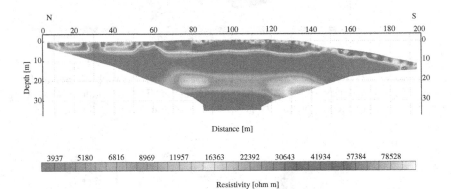

Plate 8 (Figure 8.5). DC resistivity tomography profile.

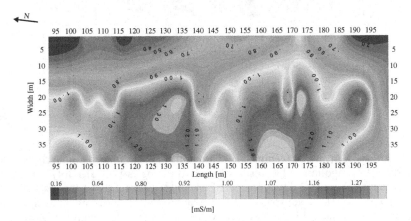

Plate 9 (Figure 8.6). Map of the EM-31 conductivity results.

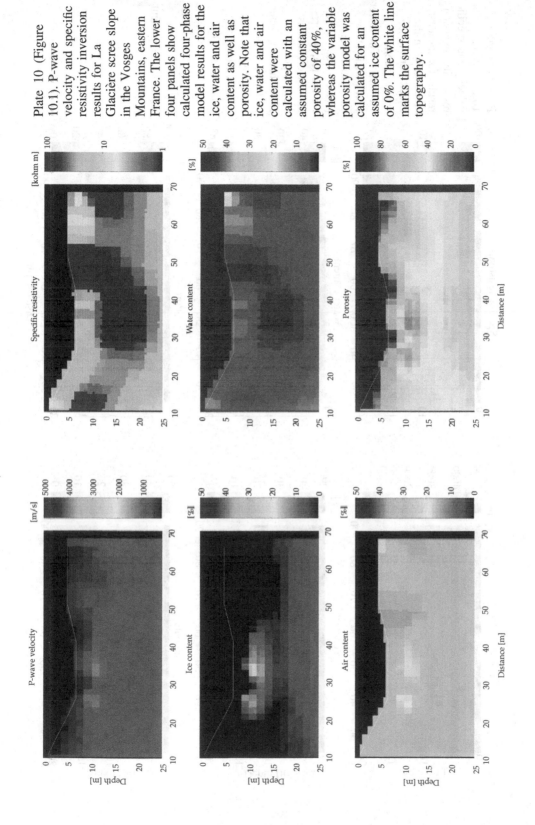

Plate 10 (Figure 10.1). P-wave velocity and specific resistivity inversion results for La Glacière scree slope in the Vosges Mountains, eastern France. The lower four panels show calculated four-phase model results for the ice, water and air content as well as porosity. Note that ice, water and air content were calculated with an assumed constant porosity of 40%, whereas the variable porosity model was calculated for an assumed ice content of 0%. The white line marks the surface topography.

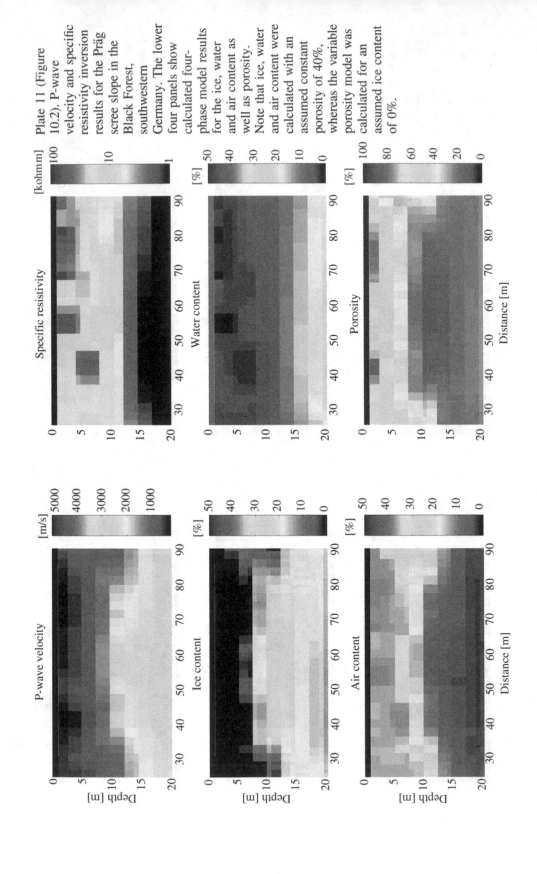

Plate 11 (Figure 10.2). P-wave velocity and specific resistivity inversion results for the Präg scree slope in the Black Forest, southwestern Germany. The lower four panels show calculated four-phase model results for the ice, water and air content as well as porosity. Note that ice, water and air content were calculated with an assumed constant porosity of 40%, whereas the variable porosity model was calculated for an assumed ice content of 0%.

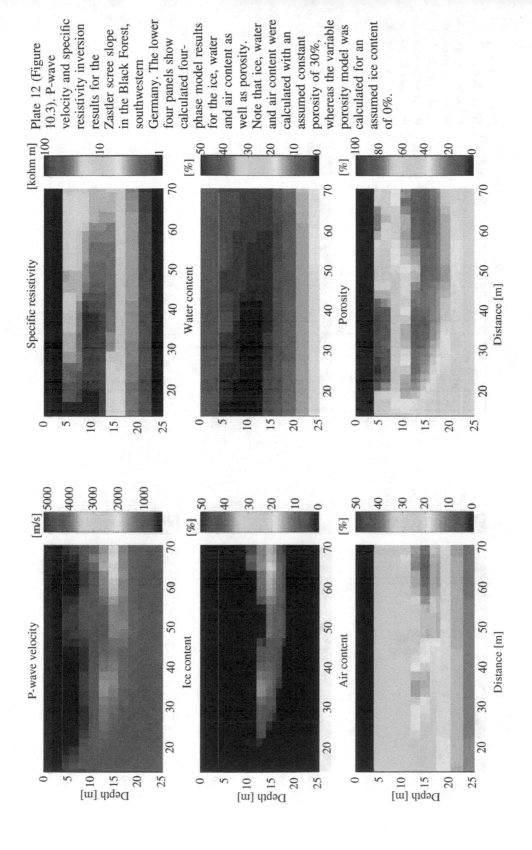

Plate 12 (Figure 10.3). P-wave velocity and specific resistivity inversion results for the Zastler scree slope in the Black Forest, southwestern Germany. The lower four panels show calculated four-phase model results for the ice, water and air content as well as porosity. Note that ice, water and air content were calculated with an assumed constant porosity of 30%, whereas the variable porosity model was calculated for an assumed ice content of 0%.

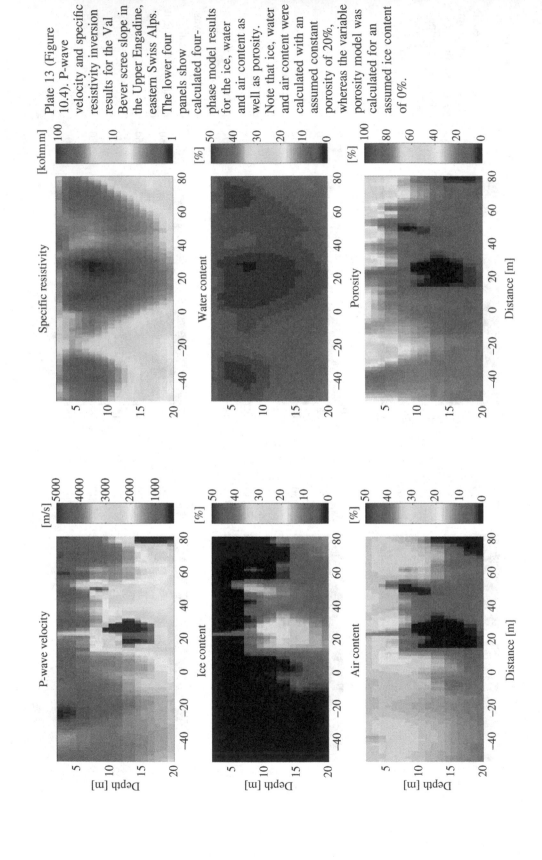

Plate 13 (Figure 10.4). P-wave velocity and specific resistivity inversion results for the Val Bever scree slope in the Upper Engadine, eastern Swiss Alps. The lower four panels show calculated four-phase model results for the ice, water and air content as well as porosity. Note that ice, water and air content were calculated with an assumed constant porosity of 20%, whereas the variable porosity model was calculated for an assumed ice content of 0%.

4

Ground-penetrating radar

I. Berthling and K. Melvold

4.1 Introduction

Ground-penetrating radar (GPR) is a geophysical method for subsurface investigation that utilises electromagnetic signals transmitted into the ground as pulses from an antenna. A receiver antenna picks up energy that is partially reflected as the signal passes through a dielectric boundary in the ground. Compared to other geophysical methods, GPR supplies data with very high vertical resolution, a potential high recording speed and real-time display of the acquired data. Commercial GPR systems have only been available since the mid 1970s and the first digitally controlled GPR system was introduced by Sensors & Software Inc. in the mid 1980s. Of early scientific applications of GPR, radar measurements in cold glacier ice are probably the most noteworthy and the technique became even more important within glaciology in the mid 1970s when technical development facilitated GPR applications also on temperate ice (see also Chapter 13). GPR was also applied early within permafrost studies (Annan and Davis 1976, Davis *et al.* 1976). GPR is today one of the standard methods for subsurface investigations, and the fundamentals of the method are provided in textbooks such as Daniels (1996) and Reynolds (1997).

The range and number of GPR applications have in general risen sharply during the past 4–5 years. This rise is also noticeable within the fields of permafrost and periglacial research. However, despite the early promising results from GPR profiling in cold environments, the absolute number of applications is still rather limited. An overview of GPR applications within periglacial, glacial and glacio-geological studies will not be provided here, as several papers have explored or reviewed the potential of the GPR method within the glacial and periglacial environment (Moorman and Michel 2000, Plewes and Hubbard 2001,

Applied Geophysics in Periglacial Environments, eds. C. Hauck and C. Kneisel. Published by Cambridge University Press. © Cambridge University Press 2008.

Berthling *et al.* 2003, Moorman *et al.* 2003, Dowdeswell and Evans 2004) as well as within related scientific fields (Moorman 2001, Overgaard and Jakobsen 2001, Neal 2004). Many further contributions and applications must be expected in the near future.

4.2 Measurement principles

The basic principle of GPR measurements is to transmit electromagnetic pulses of suitable frequency down into the ground through a transmitting antenna, and to detect the reflected energy as a function of time, amplitude and phase from any subsurface targets through a receiver antenna. If the electromagnetic travel velocity of the subsurface is known, time can be converted to depth.

As the electromagnetic waves propagate into the ground, the power decreases as the inverse square of distance. Wave reflections are generated from the boundaries of materials of different electromagnetic properties. The large contrast between the electromagnetic properties of rock, ice, water, and some sediments makes GPR a particularly effective method for mapping permafrost structures (Annan and Davis 1976). However, reflections may also originate from anisotropy due to fine layering or density variations of a single material, or from objects on the surface creating an interference pattern.

The propagation of electromagnetic signals is determined by the electromagnetic properties of the material, namely the electric permittivity (ε), the electrical conductivity (σ) and the magnetic permeability (μ). The magnetic permeability is normally assumed to be of little importance in GPR applications. The electric permittivity consists of both a real part (the dielectric constant) and an imaginary part (a loss factor). The dielectric constant is normally expressed relative (ε_r) to that of free space (ε_0), and earth materials then have values between 1 and 80. Due to the polar nature of water, the water content of a material is one important control on its dielectric constant, and measurements of dielectric constant are commonly used for determining water content of soils. Electrical conductivity may vary along much broader scales, determined largely by the amount of dissolved salts present in the water (see also Chapters 1 and 2). For low-loss materials, that is materials without substantial amounts of ions in the pore water or the clay structure, some simplified relationships between electromagnetic propagation and electrical properties may be stated. Then, the velocity of the electromagnetic wave is

$$v = c_0/(\varepsilon_r)^{1/2}, \tag{4.1}$$

where v is the electromagnetic wave velocity, c_0 is the speed of light in free space and ε_r is the relative dielectric constant. While propagating in a medium, the amplitude of the electromagnetic wave will decline as

$$A = A_0 e^{-az},\tag{4.2}$$

where a is an attenuation constant, which for a low-loss medium can be approximated as

$$a = \frac{\sigma}{2}(\mu/\varepsilon)^{1/2},\tag{4.3}$$

and z is the travel distance.

Finally, for wave propagation normal to a boundary between sufficiently thick layers, the reflection coefficient is

$$R \approx \frac{(\varepsilon_{r1})^{1/2} - (\varepsilon_{r2})^{1/2}}{(\varepsilon_{r1})^{1/2} + (\varepsilon_{r2})^{1/2}},\tag{4.4}$$

where ε_{r1} is the dielectric constant of the host medium and ε_{r2} the dielectric constant of the target (layer) medium.

For reflections to be identified, a rule of thumb states that R^2 should be larger than 0.01 (Annan and Cosway 1992). If the wavelength of the electromagnetic waves is large relative to the thickness of a layer, the situation is more complicated. Generally, the amplitude of the reflection will depend both on the thickness of the layer and on the Fresnel reflection coefficient of the medium, and the high-frequency energy will be reflected while the lower frequencies are transmitted (Annan 1998).

4.3 Data acqusition

4.3.1 GPR systems

There are two different types of GPR systems available. The most common system is the time-domain impulse radar. It consists of five components: two identical transmitter and receiver antennae (bistatic radar); a transmitter that generates a short electromagnetic pulse; a receiver that picks up the reflected pulse; a control unit that keeps track of the signal timing and performs real-time processing; and a computer for manipulating survey parameters, data storage and real-time data display. Specific radar systems may have combined the control unit and the computer, while other (shielded) systems combine all components except the computer into a single unit. Such a radar system operates on a single centre frequency and bandwidth, determined by the antennae characteristics. The alternative system is the step-frequency system, where pairs of different antennae are employed simultaneously (e.g. Kong and By 1995, Hamran *et al.* 1998). Here, continuous waves or stepped waves are transmitted and measurements of both magnitude and phase are collected at several frequencies over the frequency band of interest. This enables more information on subsurface conditions to be recorded during a single profile. The step-frequency system also has a higher signal-to-noise ratio (Hamran and Aarholt 1993). However, few commercial step-frequency systems exist, and the following treatment refers to time-domain impulse radars only.

4.3.2 Antenna set-ups

In general, there are three main types of antenna set-ups used for GPR profiling: the common offset and the common mid-point (CMP) are most frequently used. In addition, the WARR (wide-angle reflection and refraction) set-up involves keeping one antenna (either transmitter or receiver) fixed in space and moving the other away. In the CMP type of survey, the two antennae are moved apart at constant intervals. This enables estimation of the travel velocity of the medium and is often used in addition to the common-offset mode. In the common offset type, transmitter and receiver antennae are kept at a user-determined, constant distance during profiling. Choice of antenna separation may simply be done from the manufacturer's recommended minimum distance for the different antennae, or may be evaluated based on the specific targets of interest. An estimate for optimal antenna spacing may be obtained from the following relationship (Annan 1998):

$$ S = \frac{2d}{\left(\varepsilon_r - 1\right)^{1/2}}, \tag{4.5} $$

where d is depth to target and ε_r is the dielectric constant. Too close spacing will cause saturation of the signal from the direct wave, and may mask reflections from the upper parts of the profile. A large spacing of antennae causes two-way traveltimes in the upper layers to increase significantly and thus a compression of the signals in the time domain, and a decrease of depth resolution.

For common-offset profiling one has the choice of moving the antennae continuously or stepwise at predetermined intervals. In step-mode, one may better utilise the advantages of horizontal stacking, and better ground coupling is anticipated. More consistent reflections and less noise can thus be expected. However, surveying speed is reduced. When moving the antennae continuously, measurements are triggered either at specific time intervals (that can be converted to horizontal distance if the towing speed is known) or at specific distances controlled by an odometer wheel. Some manufacturers also allow triggering by a hip-chain, which in rugged terrain is a very convenient system.

Antenna orientation may also be an important choice. The most commonly used geometry is to align the antennae perpendicular to the survey line direction, parallel but not in line with each other. This ensures data that are less affected by reflections from the side of the survey line. However, if the aim of the survey is to search for and detect objects, antenna orientation parallel to the survey line will provide a broader subsurface footprint. It is also of advantage to have the antennae orientated parallel to the strike direction of any structures of interest. Finally, appropriate orientation of the antennae can be used to minimise the effects of surface reflections (van der Kruk and Slob 2004).

4.3.3 The choice of antenna frequency

There are two basic considerations that must be made when planning a GPR survey: the desired depth penetration and the resolution needed for the problem at hand. The depth penetration of a radar system is not straightforward to predict. A basic requirement is that the received reflected power from an object must be strong enough to be detectable by the system. This can be evaluated by the radar range equation, which relates the received power from a scattering object to the transmitting power, antenna gain and the distance to the object, and by the signal-to-noise ratio of the receiver. However, many of the parameters in the radar equation are generally not known. GPR system characteristics such as performance factor and antenna pattern provide basic constraints, while the electrical properties of the ground, and the character and size of the reflector in question are site-dependent constraints.

The basic decision to be made is that of antenna centre frequency. Lower frequencies generally give better depth penetration, but there is a trade-off on resolution and portability. Forward modelling, offered by some commercial GPR interpretation software packages, is a useful tool to aid in this choice. With respect to resolution, both the horizontal and the vertical resolution must be considered. The horizontal resolution is determined by wavelength and depth, so that lower frequencies and larger depths decrease the resolution. Reflectors must have a radius larger than the first Fresnel zone in order to be resolved in their lateral dimension. With a 50 MHz antenna and a medium velocity of $0.15 \, \text{m ns}^{-1}$, the wavelength is about 3 m. Minimum reflector radius r can be estimated from

$$r = (\lambda h/2 + \lambda^2/16)^{1/2}, \tag{4.6}$$

where λ is wavelength and h is depth to reflector (McQuillin *et al.* 1984). This yields a critical radius of 5.5 m at 20 m depth. Vertical resolution is the minimum distance between two reflectors so that these reflectors can be distinguished, and is determined by the wavelength and width of the reflected pulse. For a short pulse, true resolution is in the order of $\lambda/3$–$\lambda/2$ (Trabant 1984) and lower frequencies thus give lower resolution. The centre frequency should also be chosen so as to reduce chaotic reflections from material heterogeneity in soils and rock at a smaller scale than what the survey seeks to explore. The typical dimension of such potential unwanted reflectors should be much shorter than the signal wavelength, that is much shorter than v/f, where v is the velocity of the medium and f is the centre frequency. Annan (1998) states that a factor of 10 is appropriate.

4.3.4 *User-controlled parameter settings*

Depending on the specific GPR system used, a number of parameters may be determined by the user to optimise data collection. The *time window* determines how long the system will record signals from the receiver antenna after a pulse has left the transmitter antenna. The necessary time window can be found by considering the maximum depth and the minimum velocity likely to be encountered. The *sampling interval* determines how often the signal that is received is measured, and thus how well its form is represented. The frequency must be at least twice as high as the highest frequency of the signal, in order to reproduce this correctly. Half the sampling frequency is termed the Nyquist frequency (f_N). If a specific frequency in the signal is Δf higher than f_N, then this frequency will be reproduced as $f_N - \Delta f$. This is commonly termed aliasing. GPR antennae transmit signals with frequencies in approximately the interval $\langle 0.5f, 1.5f \rangle$ where f is the centre frequency of the antennae. With a factor of safety of two, the sampling frequency should be at least six times the centre frequency of the antennae, to be able to get real vertical resolution.

Similar considerations have to be made concerning the *station spacing*: the horizontal distance between discrete radar measurements. This distance must not be too large, or steeply dipping reflectors will not be imaged properly. In this case, the critical maximum distance is (e.g. Annan 1998)

$$\Delta_x = \frac{c_0}{4f(\varepsilon_r)^{1/2}} = \frac{1}{4f}v = \frac{\lambda}{4}. \qquad (4.7)$$

In areas of more horizontally aligned reflectors, this criterion can be compromised. During data acquisition, individual measurements may be stacked. *Vertical stacking* or *multifold coverage* is a combination of profiling and CMP. With *horizontal stacking*, the GPR systems perform a user-determined number of measurements at a single station and then average these traces. Stacking will in general improve the signal-to-noise ratio, but a large number of stacks may also introduce noise if the GPR system is moved continuously over a non-homogeneous subsurface. Horizontal stacking may also be performed as a post-processing step by averaging adjacent traces.

4.3.5 Survey geometry

Most GPR profiles utilise 2D radar reflection profiles of the subsurface. During common-offset surveying, antennae are either moved continuously along the ground or moved stepwise at constant intervals. In the first case, signals are triggered either at regular time intervals, approximately corresponding to the desired spatial sampling interval between traces (the step-size) or at fixed distances controlled by a distance wheel or a hip-chain. On rough terrain, accurate spacing of measurement points is very difficult to obtain. In the second case the step-size determines signal triggering directly and the antennae are stationary during a measurement. Although the latter method has definite advantages with respect to antenna/ground coupling and trace stacking (Annan and Cosway 1992), the need for surveying speed usually requires continuous profiling. Some GPR systems enable the user to tick-mark locations or special features during continuous profiling. This may aid in positioning of the profile. Ground-plan positioning of the survey line is obviously necessary, but also height information is desirable. This enables height correction of the GPR profile, which eases interpretation significantly. This terrain correction is applied as a last step in data processing (Figure 4.1).

Sometimes, 3D surveys may be necessary or desirable. A true 3D survey involves very close spacing of transect lines, so that data from individual traces overlap. This necessitates accurate positioning of data points in ground plan and height. In pseudo 3D surveys, data are collected so that surveying lines intersect and form a regular or irregular grid.

4.4 Data processing

Processing of GPR data involves techniques that on the one hand seek to enhance any reflections by amplifying and filtering, and on the other hand seek to represent the geometry of reflections more correctly. Some of these techniques may be applied in real-time, so that the sub-surface is more readily visualised during data collection, but the original data are normally (and should definitely be) stored unaltered. Some processing may be performed within the software supplied with the GPR system, but purchase of a designated GPR/seismic processing software package is recommended. Here, we provide a short overview of the main processing steps (Figure 4.1). Neal (2004) provides an excellent, more in-depth treatment of this matter.

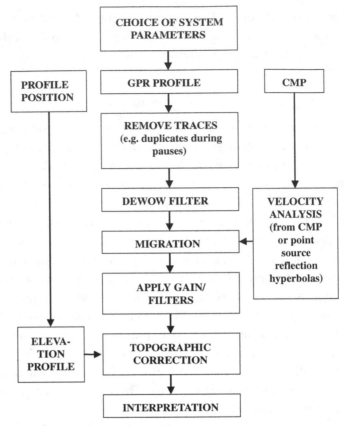

Figure 4.1. Flow chart showing the most important steps in GPR data acquisition and processing. After processing and initial data interpretation, there may be a need for adjusting some of the processing steps to remove processing artefacts, increase visibility of special features etc.

4.4.1 Horizontal stacking, dewow, gain and filtering techniques

Horizontal stacking is often performed during data collection, but adjacent traces may also be averaged as a first processing step to remove or reduce ambient noise. The second step is usually to apply a 'dewow' filter which removes very-low-frequency components of the data, a peculiarity of GPR systems. The third step seeks to compensate for the rapid attenuation of the radar signal with depth. This involves applying some kind of 'gain' to the data. Sometimes, a constant gain may be sufficient and all parts of the trace are then amplified by the same amount. More often, the application of a time-dependent gain is necessary, to enable weak signals from deep targets to be recognised simultaneously with greater amplitude signals from the shallow subsurface. This 'time gain' seeks to equalise the amplitudes along the trace according to the rapid fall-off of signal strength with depth. However, as the attenuation is a function of both depth and ground properties, application of a time gain is not straightforward. The spherical and exponential compensation (SEC) gain attempts to compensate for the geometric spreading and exponential dissipation of signal energy. This gain maintains some fidelity with respect to the relative amplitude of the reflections. If one wishes to see 'all' information, irrespective of relative amplitudes, an automatic gain control (AGC) may be used. The AGC computes an average signal over a (user-defined) time window, and amplifies or attenuates the data point in the centre of this time window by the ratio of the desired output value to the average signal amplitude. Although the AGC will normally enhance signals, it also typically creates low-amplitude zones below strong signals. The choice of time window is therefore important.

An optional fourth processing step is to apply filtering procedures on the data. These may be performed both in the time domain (along a trace) or in space (between neighbouring traces). Such procedures may increase the signal-to-noise ratio or may be used to remove spikes in the data. An example of a spatial filtering technique is background subtraction, which may remove time-synchronous system artefacts and ringing.

4.4.2 *Migration*

The electromagnetic wave propagates into the ground as a section of a sphere, with increasing radius with depth. This causes the GPR image of point reflectors and sloping reflectors to be systematically distorted, as a specific return signal may come from anywhere on the spherical wavefront. Point reflectors are continuously imaged from some time before to some time after they are directly beneath (or perpendicular to the side of) the surveying line. This causes point reflectors to have a hyperbola structure in the GPR image. For the same reason, sloping reflectors are imaged with an apparent dip less than the real slope. Migration routines seek to compensate for these effects. Migration is a very important processing step which requires good knowledge of the subsurface velocity. It is also very dependent on subjective operator choices and a bias towards preconceived ideas of profile appearance is likely. In areas where topographic variations are in the same order as the penetration depth of the radar signal, standard migration routines are not appropriate and topographic migration algorithms have to be used (Lehmann *et al.* 1998, Lehmann and Green 2000). After a profile has been migrated, topographic corrections can be applied based on the elevation along the GPR profile.

Examples of the influence of processing steps, including migration, on the appearance of GPR profiles are provided by Neal (2004).

4.4.3 Two-way traveltime and depth conversion

The GPR data are recorded in the time domain, and the 'depth' axis in a GPR profile is the two-way traveltime (TWT). A conversion of TWT to depth is obviously desirable, but this is dependent on knowledge of subsurface velocities along the profile. Often, some average velocity can be assumed, but one should be aware that the profile will be visually distorted if the electromagnetic velocity changes with depth, because the depth scale will expand or contract. A simple example is that of permafrost terrain, where the signal velocity in the active layer may be much lower than within the permafrost. The active layer will then be displayed much thicker than its real depth if the signal velocity within the permafrost is used for depth conversion. In such cases, a variable vertical scale can be used. Similar changes may also occur laterally, underlining the need for CMP surveys alongside common-offset measurements.

4.4.4 Concluding remarks

During processing and the following interpretation of the profile, reflections must be treated with care with respect to their origin. First of all, the upper part of the profile is obscured by the direct air and ground waves. Second, reflections found in a profile may not only come from interfaces between different ground materials. One must consider possible reflections from the ground surface, heterogeneity within a single ground material, thin layers, secondary events, noise, point reflections from buried boulders etc. Quantitative analysis of the number of layers and the detailed structure and composition of the subsurface must therefore involve a detailed analysis of the radar signals, not only a visual interpretation of the GPR image.

4.5 Periglacial applications and particularities

GPR applications in mountainous terrain face some special considerations, depending on the type of equipment at hand, the object under investigation and the time of the year. With respect to equipment, we will focus on GPR systems where antennae, transmitter and receiver electronics, the control unit (radar console) and the computer are separated. Such systems, like pulseEKKO 100 from Sensors & Software Inc. (now replaced by the pulseEKKO PRO) and RAMAC GPR from Malå Geoscience, offer large user flexibility. However, the number of components may introduce logistic challenges.

4.5.1 Set-up of the components

In mountainous terrain, there are generally two options for continuous profiling. During winter or generally on ice and snow, the equipment can be mounted on a wooden sledge and dragged by hand or snowmobile across the terrain (see Figure 13.2). In summer, a rough terrain surface generally requires the equipment to be carried along the profile. Two or three people are then required to carry the set-up, depending on the antennae size, which increases with lower centre frequency. Both options require some consideration with respect to connections between the different system components. Transmitter and receiver electronics, including batteries, are somewhat heavy, and should be strapped to an external frame or at least to the antennae to remain stable during profiling. This is important both to prevent wear of the antenna/electronics socket jacks and to prevent disruption of the connections between receiver/transmitter electronics and the radar console during profiling. Transmitter and receiver are connected to the radar console by fibre-optical cables, while the radar console and the computer are linked through the RS232 and/or the parallel port of the computer. In addition, the console and the computer might need to be connected to an external 12 V battery. Ideally, unshielded computers and the radar console should be kept at some distance from the antennae.

The fibre-optical cable connections are a weak link in this set-up. When the GPR system is carried, one person carries and controls the radar console, computer and batteries while one or two people carry the antennae and receiver/transmitter electronics. This requires some flexibility, and it is necessary to introduce a cord anchorage, for instance in the form of a rope that is slightly shorter than the fibre-optical cables, to avoid disruption of connections. The fibre-optical cables are very sensitive, and care is needed to avoid them being trampled on or torn across boulders etc.

Under winter conditions, the set-up itself may be easier, as the main part of the equipment can be mounted on a sledge. However, if it is necessary or favourable to view the data while profiling, one may be faced with the same problem of having to keep radar console and computer separated from the antennae. One possible solution to this problem is to employ a computer with an external screen and keyboard (Figure 4.2), but a deterioration of the signal returns can be expected when the radar console is close to the antennae (see Chapter 16). For the pulseEKKO 100 system, the manufacturer recommends a distance of at least 5 m between the receiver and console/computer. Some GPR systems use shielded antennae, and this problem is then eliminated.

4.5.2 Temperature

In addition, coldness requires special considerations. If profiling starts only a short time after the GPR system is exposed to markedly colder (or warmer)

Figure 4.2.   A pulseEKKO 100 GPR system used on river ice. The radar console, one or two 12 V batteries and a rugged computer are stored and connected inside the box (upper picture), and connected by a 5 m cable to an external display with heating possibilities and keyboard (lower picture). 100 MHz and 200 MHz antennae are fixed at 1 m and 0.5 m distance on the sledge, and the only connections that must be made in the field are to mount the antennae electronics on the antennae, and the fibre-optical cables to the transmitter and receiver. Such a system enables better productivity during fieldwork but introduces some noise on the radar profile (see Chapter 16).

ambient temperatures, time-zero drift can be anticipated (see Section 2.3.4). The electronics should therefore be allowed to adjust to ambient temperatures before measurements are started. During winter, the assemblage of the radar system is a chilly task, and it is of definite advantage to prepare a system where all possible connections can be put together beforehand (Figure 4.2). The fibre-optical cables should be well marked with their correct connection. Battery capacity may also be a significant issue. Computers will generally need external power from a 12 V battery, and during extreme cold a modern rugged field computer with self-heating capabilities is often a requirement. Also the display may have to be specially adjusted, as the sharp light during winter makes it extremely difficult to view a traditional LCD display. Modern field computers have transflective LCD displays that reduce this problem.

4.5.3 Depth penetration

In permafrost environments, measurements should be performed during winter if deep penetration of the radar signal has first priority. Although an extra layer (snow) is introduced, which may cause some problems with interpretation of the upper layers, signal attenuation because of a melted active layer will be avoided. However, experience from rock glaciers on Svalbard indicates that the importance of this effect will depend on local circumstances (see Chapter 12).

4.6 Recommendations

4.6.1 Survey preparation

When the decision to use GPR for a specific survey has been made, there are a number of issues to consider. The GPR equipment at hand puts some basic constraints on the possibilities. The commercial systems described above offer the flexibility of choosing among a range of antennae and survey modes. Other systems exist that offer easy handling, by combining shielded antennae and electronics into a robust, small unit. In this case, there are fewer decisions to be made but also fewer possibilities for tailor-made survey designs. For general earth science applications one may consider renting antennae to supplement the available frequencies, so the field efforts of gathering some extra profiles are lessened when first on the site.

A combination of 50, 100 and 200 MHz antennae works well in most cases. For deeper penetration or a medium with higher loss, 25 MHz antennae can be used and some manufacturers now also offer 12.5 MHz antennae. Low-frequency antennae are long and not too versatile in rough terrain, but at least periglacial areas mostly lack trees! Malå Geoscience offer a solution to this problem, with 50 MHz antennae mounted after one another in a rough, flexible casing that is dragged behind the operator. Antennae of 500 MHz or even higher frequency are necessary for snow surveys (see Chapter 15) and could be useful also for studies of the active layer and ice conditions.

A point that should be stressed with respect to preparations before fieldwork is always to bring enough spare parts and batteries. The fibre-optical cables are especially vulnerable, and although repair is possible replacement is so much easier. As stated above, prepare the computer for connection to a 12 V power source. Design a system for carrying or dragging all equipment along the profile, and check this set-up carefully in a field setting before taking the equipment to a remote location. And, in case you have purchased a brand-new rugged laptop to use with an old GPR system, check compatibility of the GPR software!

In most cases it is possible to determine user-controlled parameter settings beforehand for the different antennae and the specific surveying purpose at hand. Sometimes, full sets of parameters can be stored as specific files, so that each parameter does not have to be entered manually before a GPR profile is gathered. The considerations to be made regarding these parameters are discussed above. For the time window, it is recommended to choose a value that enables deeper returns than one expects, as the increase in file size is irrelevant for current computers. For the same reason, the sampling frequency should be higher and the spatial sampling interval lower than the minimum criteria stated above. Typical values are listed in Table 4.1.

4.6.2 *The GPR survey*

Common-offset profiling with continuously moving antennae is probably the most frequently used data acquisition mode. However, as the medium velocity is an important parameter for migration, interpretation and display of GPR data, one or several CMP profiles should be collected as a supplement to the common-offset profiles. Antenna spacing should be at least 0.5 m (200 MHz), 1 m (100 MHz) and 2 m (50 MHz), and the default antenna orientation is perpendicular to the survey line, with the transmitter closest to the survey direction. Differential GPS is the most accurate and time-saving method for determining the geometry of the survey line. Modern GPR systems offer integration of GPR and GPS data, while for older systems some post-processing is necessary. If surface reflectors are present, one should measure their position relative to the profiles. This makes it much easier to recognise these unwanted reflectors in the radar profile.

Available time and the level of detail needed for the survey determine the spacing of the profiles. A network of surveying lines (pseudo 3D) is of course advantageous because the spatial extent of single reflectors and reflector patterns can be better resolved.

Table 4.1. *System set-up parameters and values for the dielectric constant, wavelength, velocity and approximate vertical resolution, defined for the most common antennae and some earth materials*

Antenna centre frequency (MHz)	Maximum sampling interval (ns)	Dielectric constant (ε_r)	Wavelength (m)	Maximum station spacing (m)	Vertical resolution (m), ($\lambda/2$)	Velocity (m/ns)	Material
50	3.3	4–8	3.0–2.1	0.75–0.53	1.5–1.1	0.15–0.11	Permafrost[a]
		25	1.2	0.30	0.60	0.060	Active layer[b]
		80	0.67	0.17	0.34	0.034	Water
100	1.67	4–8	1.5–1.1	0.37–0.27	0.75–0.53	0.15–0.11	Permafrost[a]
		25	0.6	0.15	0.30	0.060	Active layer[b]
		80	0.34	0.08	0.17	0.034	Water
200	0.83	4–8	0.75–0.53	0.19–0.13	0.37–0.27	0.15–0.11	Permafrost[a]
		25	0.30	0.07	0.15	0.060	Active layer[b]
		80	0.17	0.04	0.08	0.034	Water

[a] The indicated range for permafrost materials is very approximate, dielectric constant values will probably often be in the lower range. Pure ice has a dielectric constant of 3–4.
[b] The range for active layer properties is even higher. Here, only an example is provided.

4.6.3 Post-processing and interpretation

Important processing steps are displayed in Figure 4.1. The possibilities may be constrained by the software at hand. Migration is a very important processing step, improving the geometrical representation of subsurface structures. Unmigrated data have potential limitations that must be taken into account. The application of different types of gains and filters are typical procedures where one may play around to produce an image that looks visually attractive. However, during these processing steps it is very important that the operator keeps in mind what features are the focus of the study, so that the 'right' reflectors are highlighted. A basic choice is whether one should keep information on relative amplitudes (using a time gain), or equalise signals (using an automatic gain control). Whatever choice is made, it is vital to keep in mind what is lost and what is gained for the method used.

When post-processing is finished and a radar profile can be visually inspected, it is important to keep the limitations in mind. Do not automatically equate a reflector with a real layer or feature in the subsurface.

4.6.4 Final remarks

If you in some way or another can borrow, rent or buy a GPR system, just do it! There are many features in the periglacial and glacial realm that still await the application of GPR, and several others where far more data are needed.

REFERENCES

Annan, A. P. (1998). *Ground Penetrating Radar Workshop Notes*, Sensors and Software Inc. Mississauga, Ontario.

Annan, A. P. and Davis, J. L. (1976). Impulse radar sounding in permafrost. *Radio Science*, **11**, 383–394.

Annan, A. P. and Cosway, S. W. (1992). Ground penetrating radar survey design. *Proceedings of the Symposium on the Application of Geophysics to Engineering and Environmental Problems (SAGEEP)*, Oakbrook, USA, 329–351.

Berthling, I., Etzelmüller, B., Wåle, M. and Sollid, J. L. (2003). Use of Ground Penetrating Radar (GPR) soundings for investigating internal structures in rock glaciers. Examples from Prins Karls Forland, Svalbard. *Zeitschrift für Geomorphologie, Supplement*, **132**, 103–121.

Daniels, D. J. (1996). Surface-penetrating radar. *Electronics & Communication Egineering Journal*, **8**(4), 165–182.

Davis, J. L., Scott, W. J., Morey, R. M. and Annan, A. P. (1976). Impulse radar experiments on permafrost near Tuktoyaktuk, Northwest Territories. *Canadian Journal of Earth Sciences*, **13**, 1584–1590.

Dowdeswell, J. A. and Evans, S. (2004). Investigations of the form and flow of ice sheets and glaciers using radio-echo sounding. *Reports on Progress in Physics*, **67**, 1821–1861.

Hamran, S.-E. and Aarholt, E. (1993). Glacier study using wavenumber domain synthetic aperture radar. *Radio Science*, **28**(4), 559–570.

Hamran, S.-E., Erlingsson, B., Gjessing, Y. and Mo, P. (1998). Estimate of the subglacier dielectric constant of an ice shelf using a ground-penetrating step-frequency radar. *IEEE Transactions on Geoscience and Remote Sensing*, **36**, 518–525.

Kong, F. N. and By, T. L. (1995). Performance of a GPR system which uses step frequency signals. *Journal of Applied Geophysics*, **33**, 15–26.

Lehmann, F. and Green, A. G. (2000). Topographic migration of georadar data: implications for acquisition and processing. *Geophysics*, **65**, 836–848.

Lehmann, F., Vonder Mühll, D., van der Veen, M., Wild, P. and Green, A. (1998). True topographic 2-D migration of georadar data. *Proceedings of the Symposium on the Application of Geophysics to Environmental and Engineering Problems (SAGEEP)*, Chicago, 107–114.

McQuillin, R., Bacon, M. and Barclay, W. (1984). *An Introduction to Seismic Interpretation*. Graham & Trotman Ltd.

Moorman, B. J. (2001). Ground-penetrating radar applications in paleolimnology. In *Tracking Environmental Change Using Lake Sediments: Physical and Chemical Techniques*, ed. Last, J. P., Kluwer Academic Publishers, pp. 23–47.

Moorman, B. J. and Michel, F. A. (2000). Glacial hydrological system characterization using ground-penetrating radar. *Hydrological Processes*, **14**(15), 2645–2667.

Moorman, B. J., Robinson, S. D. and Burgess, M. M. (2003). Imaging periglacial conditions with ground-penetrating radar. *Permafrost and Periglacial Processes*, **14**(4), 319–329.

Neal, A. (2004). Ground-penetrating radar and its use in sedimentology: principles, problems and progress. *Earth-Science Reviews*, **66**(3–4), 261–340.

Overgaard, T. and Jakobsen, P. R. (2001). Mapping of glaciotectonic deformation in an ice marginal environment with ground penetrating radar. *Journal of Applied Geophysics*, **47**(3–4), 191–197.

Plewes, L. A. and Hubbard, B. (2001). A review of the use of radio-echo sounding in glaciology. *Progress in Physical Geography*, **25**(2), 203–236.

Reynolds, J. M. (1997). *An Introduction to Applied and Environmental Geophysics*. John Wiley & Sons.

Trabant, P. K. (1984). *Applied High-resolution Geophysical Methods*. International Human Resources Development Corp., Boston, 265pp.

van der Kruk, J. and Slob, E. C. (2004). Reduction of reflections from above surface objects in GPR data. *Journal of Applied Geophysics*, **55**, 271–278.

Part II

Case studies

5

Typology of vertical electrical soundings for permafrost/ground ice investigation in the forefields of small alpine glaciers

R. Delaloye and C. Lambiel

5.1 Introduction

During the Little Ice Age (LIA), many small glaciers – usually less than 1 km wide and some having completely vanished since that time – overlaid permafrost areas in the Alps above approximately 2500 m a.s.l. Strong mechanical (e.g. push moraines) and thermal disturbances (e.g. permafrost degradation) of the former frozen sediments occurred. LIA glacier forefields located in the discontinuous belt of permafrost are thus complex geomorphic features (Figure 5.1) including various types of ground ice and coalescent frozen and unfrozen ground conditions (Evin and Assier 1983). Since the 1980s, several studies have been carried out to map the ground ice distribution in such recently deglaciated terrains (e.g. Evin 1992, Kneisel 1999, 2003a). From the same perspective, we performed about 100 vertical electrical soundings (VES) on thirteen sites in the western Swiss Alps and the Pyrenees between 1997 and 2003 (Delaloye and Devaud 2000, Delaloye *et al.* 2003a,b, Reynard *et al.* 2003, Delaloye 2004, Lambiel *et al.* 2004, Lugon *et al.* 2004, Lambiel 2006). The following case study proposes an interpretative typology of VES measured in this kind of glacial/periglacial environment. The typology is complemented by additional indications on the ground surface thermal regime.

5.2 Method

The VES technique (see Chapter 1) is logistically adapted to permafrost investigations in difficult terrain and remote areas. The power supply requirement is limited and, in contrast to the ERT method, the basic equipment remains lightweight (and not expensive!). In spite of the blocky nature of the ground surface (replacing metallic electrodes with sponges soaked in salt water ensures, when

Applied Geophysics in Periglacial Environments, eds. C. Hauck and C. Kneisel. Published by Cambridge University Press. © Cambridge University Press 2008.

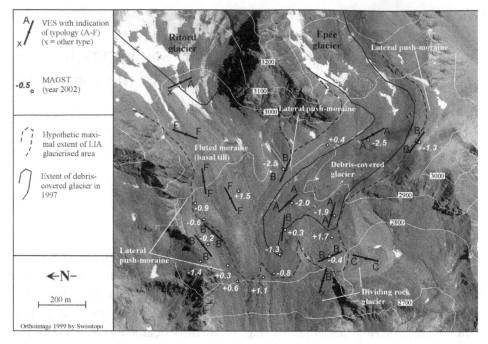

Figure 5.1. The Ritord glacier forefield, with some typical landforms, the interpretation of VES and MAGST observed in 2002. At Ritord, MAGST in 2002 is close to the mean of the period 1997–2006.

necessary, a good connection with the ground surface), apparently coherent VES results are quite easy to obtain in glacier forefields and surrounding areas. Data interpretation may be conversely more hazardous. A source of misinterpretation is the potentially strong lateral subsurface heterogeneity of the ground properties. It can be prevented by systematically applying an asymmetrical Hummel configuration, which permits one to detect the changes in resistivity towards both sides of the profile (Vonder Mühll 1993). Delaloye (2004) described in detail the strengths and limits of the interpretation of dissymmetrical VES in heterogeneous mountain periglacial terrain.

VES (as any other electrical resistivity data) interpretation gains in reliability when it is supported by additional geophysical data or ground surface temperature measurements. Apart from ground temperatures measured in boreholes, two further parameters of the ground thermal condition are often determined: the winter equilibrium temperature (WEqT) (close to the classical BTS – bottom temperature of the winter snow cover – value measured by probing) and the mean annual ground surface temperature (MAGST). Such parameters are however subject to significant annual changes depending particularly on both the timing and the development of the snow cover (e.g. Delaloye and Monbaron 2003, Delaloye 2004). Recently, 2D electrical resistivity tomography (ERT) has tended to replace 1D VES for permafrost investigation, e.g. in glacier forefield environments (Kneisel 2003b, 2004, Marescot *et al.* 2003, Reynard *et al.* 2003). ERT provides high-quality data that are nevertheless limited by the depth to which the inversion model provides reliable data, the so-called DOI (depth of investigation index, Marescot *et al.* 2003) and the difficulty of inverting highly contrasted resistivity values (see Chapter 1). Glacier forefields are often quite large areas and there is consequently a tendency to install long ERT profiles with large electrode spacings, which do not resolve the active layer zone (Reynard *et al.* 2003). In these cases additional shallow VES may be performed along the ERT profile for complementing and improving the ERT data (and/or additional ERT surveys with shorter spacing).

5.3 Typology

Most of the VES that were carried out in glacier forefields and their immediate surroundings can be grouped in six main types, according to the shape of the VES curve (Figure 5.2), the characteristics of the active layer, the resistive subjacent layer (as indicated by the apparent resistivity maximum, ρ_a max, and the resistivity ρ) and the ground surface thermal parameters. They are described hereafter and summarised in Table 5.1. Ranges given for the resistivity values are indicative; in particular they may differ depending on the ground lithology. The thermal parameters mentioned in the typology are based on mean values observed between 1997 and 2003.

5.3.1 Type A

The sounding curve of a VES belonging to this group comprises an initial oversteepening and a maximum value ranging between 100 and more than 1000 kΩ m. It indicates the occurrence of massive ice (often of glacial origin) very close to the surface. WEqT has been measured between $-1\,°C$ and $-8\,°C$. MAGST is generally negative and tends to be colder than in the surroundings. Summer temperatures can remain colder than $+5\,°C$. Type A was mainly obtained on debris-covered glaciers, on buried ice patches and on the internal side of lateral push moraines.

5.3.2 Type B

Type B is a decreasing continuum of type A without oversteepening: the mantle of unfrozen sediments probably exceeds 2 m in thickness. The specific resistivity

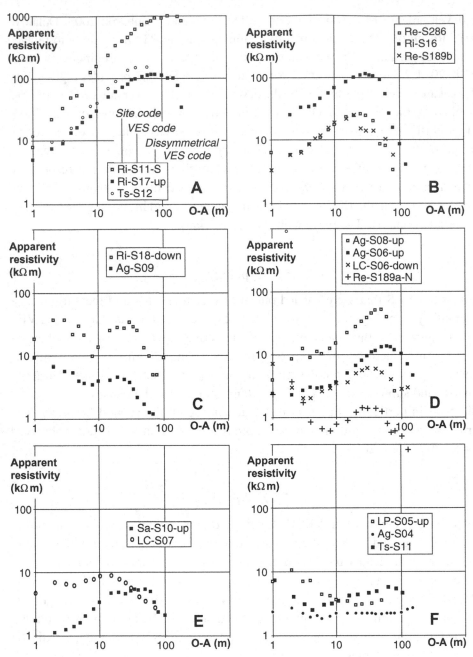

Figure 5.2. Typology of VES performed in the forefield of small alpine and pyrenean glaciers (after Delaloye and Devaud 2000). Ag: Aget (Delaloye and Devaud 2000), LC: La Chaux (Reynard *et al.* 2003), LP: La Paùl (Lugon *et al.* 2004), Re: Réchy (Delaloye 2004), Ri: Ritord (Delaloye 2004), Ts: Tsarmine (Lambiel *et al.* 2004).

Table 5.1. *VES typology and thermal characteristics*

Type	Active layer	ρ_a max	ρ max	WEqT	MAGST
A	shallow, often < 2 m	100–>1000	>1000	various, rather cold	<0 °C, colder than surroundings
B	>2 m	50–300	100–500	cold (<−3 °C)	slightly negative, approx. 1 °C warmer than type A
C	>3 m	10–>100	50–500	between −1 and −5 °C	various, depending on ground nature
D	>5 m	2–100	20–200	between 0 and −2 °C	close to 0 °C or slightly warmer
E	2–3 m	5–20	10–50	between −0.5 and −2 °C	close to 0 °C
F	no active layer	1–5		close to 0 °C	>+1 °C

ρ_a max is the maximal apparent resistivity for a VES, ρ max is the maximal specific resistivity of the frozen subjacent layer.

of the resistive subjacent layer most often ranges between 100 and 500 kΩ m. WEqT is cold (<3 °C) and MAGST is often warmer (by approximately 1 °C) than for type A, remaining nevertheless very slightly negative (−0.5/−1 °C). Type B is interpreted as a cold and thick layer of frozen sediment, in which the occurrence of massive ice of glacial origin cannot be excluded. Such VES were primarily obtained in the external part of push moraines and, sometimes, in the upper part of rock glaciers directly connected to a proglacial margin.

5.3.3 Type C

In most cases, the apparent resistivity ρ_a is relatively high for small electrode spacings due to the presence of a superficial blocky layer without fine matrix. For increasing spacings, ρ_a decreases at first indicating a thick active layer (approximately 5 m), before drawing a 'bell'-shaped curve. The layer causing the rise in ρ_a has a specific resistivity of about 10 to more than 100 kΩ m and is often not very thick (10–20 m). WEqT (between −1 °C and approximately −5 °C) as well as MAGST can vary depending on both the surface composition and the thermal state of subjacent permafrost. Type C is typically measured on blocky rock glaciers. It indicates the occurrence of frozen sediment, without any layer of massive ice close to the surface, which was not covered by a glacier during the LIA. Elsewhere, the blocky surface layer would be absent or filled with fine moraine matrix.

5.3.4 Type D

The superficial blocky layer is lacking and a significant increase in ρ_a first starts beyond AB/2 distances of 8–10 m (A and B being the two outer (current) electrodes). The specific resistivities of the deep-lying (frozen) materials range from 20 to more than 200 kΩ m and their thickness can be larger than 20 m. WEqT often ranges between 0 and $-2\,°$C; MAGST is approximately 0 °C or slightly positive. VES of type D, such as those shown in Figure 5.2, were systematically measured in zones covered by LIA glaciers that were not heavily charged with debris, as confirmed by the generalised occurrence of subglacial till. The VES curves often show that the thickness of the unfrozen surface layer exceeds 5 m. The complete freezing of the active layer in winter is uncertain. In our sites, type D appears to indicate either the degradation of former permafrost by a warm-based LIA glacier, or the current thermal degradation of permafrost pushed by a LIA glacier towards a location unfavourable to its preservation.

5.3.5 Type E

Such VES were measured in areas that have not been glacier covered for a few years or decades. ρ_a rises beyond AB/2 = 3–4 m, indicating the occurrence of a resistive layer around 2–3 m below the surface. Its specific resistivity is about 10 to 50 kΩ m; WEqT is generally warm, between -0.5 and $-2\,°$C. MAGST is close to 0 °C. Because of the shallow depth of the resistive layer, type E may be interpreted as either neo-permafrost – still in formation – or as subglacial permafrost originating when the glacier was only a few metres thick. However, according to the relatively warm surface conditions, such VES could also reflect coarse debris covered with fine material.

5.3.6 Type F

VES of type F were measured in areas covered by a LIA glacier. ρ_a is relatively constant at a low value. WEqT is usually close to 0 °C. MAGST is generally warmer than $+1\,°$C. Type F, associated with such warm ground surface conditions, indicates the absence of permafrost.

Except VES of type E that appear to be relatively rare, the other types of VES are potentially common in the proglacial margins of small alpine glaciers (Table 5.2). They do not however occur in every glacier forefield. As an example Figure 5.1 shows a glacier forefield where types D and E are lacking. Only 14% of the VES carried out on our different sites cannot be attributed to one of the main types. Their

Table 5.2. *Absolute and relative number of VES by type*

	Total	Type A	Type B	Type C	Type D	Type E	Type F	Other
Number	105	17	15	13	15	3.5	26.5	15
Percentage	100	16	14	12	14	3	25	14

A value of 0.5 is attributed to each part of the dissymmetrical VES.

interpretation, even complemented by thermal and geomorphological data, some-times remains difficult.

5.4 Conclusions

Numerous field investigations have shown that the pattern of the spatial distribution of permafrost and ground ice is somewhat similar in the forefield of every small LIA alpine glacier located in the discontinuous permafrost belt: (a) permafrost and ground ice are often restricted to the margins of the former glacierised area; (b) the central part of LIA glacier forefields may be still occupied by a degrading debris-covered glacier; (c) when not, this central area is mostly not underlain by permafrost. The proposed VES typology implies that some kinds of similar permafrost and ground ice characteristics are recurrent. In many cases, they can be identified by electrical resistivity measurements, VES in particular, corroborated by ground surface temperature measurements.

REFERENCES

Delaloye, R. (2004). *Contribution à l'Etude du Pergélisol de Montagne en Zone Marginale*. Thèse. Faculté de Sciences, Université Fribourg, Geofocus, Vol. 10.

Delaloye, R. and Devaud, G. (2000). La distribution du pergélisol dans les marges proglaciaires des glaciers de Challand, d'Aget et du Sanetschhorn (Valais, Alpes suisses). In *Beiträge zur Geomorphologie*, eds. Hegg, Ch. and Vonder Mühll, D., pp. 89–96.

Delaloye, R. and Monbaron, M. (2003). Snow effects on recent shifts (1998–2002) in mean annual ground surface temperature at alpine permafrost sites in the western Swiss Alps. *Proceedings of the 8th International Conference on Permafrost*, Zürich, Switzerland, Extended Abstracts Reporting Current Research and New Information, 23–24.

Delaloye, R., Lugon, R., Lambiel, C. and Reynard, E. (2003a). Réponse du pergélisol à l'avancée glaciaire du Petit Age Glaciaire: quelques exemples alpins et pyrénéens. *Environnements périglaciaires*, Bulletin de l'Association Française du Périglaciaire, **10**.

Delaloye, R., Métrailler, S. and Lugon, R. (2003b). Evolution du pergélisol dans les complexes glacier/glacier rocheux des Becs-de-Bosson et de Lona (VS). *Bulletin de la Murithienne, Societé Valaisanne des Sciences Naturelles*, **121**, 7–20.

Evin, M. (1992). Une moraine de refoulement au Viso (Italie). *Zeitschrift für Gletscherkunde und Glazialgeologie*, **27/28**, 11–24.

Evin, M. and Assier, A. (1983). Glacier et glaciers rocheux dans le Haut-vallon du Loup (Haute-Ubaye, Alpes du Sud, France). *Zeitschrift für Gletscherkunde und Glazialgeologie*, **19**(1), 27–41.

Kneisel, C. (1999). Permafrost in Gletschervorfeldern – Eine vergleichende Untersuchung in den Ostschweizer Alpen und Nordschweden. *Trierer Geographische Studien*, **22**, 156pp.

Kneisel, C. (2003a). Permafrost in recently deglaciated glacier forefields – measurements and observations in the eastern Swiss Alps and northern Sweden. *Zeitschrift für Geomorphologie*, **47**, 289–305.

Kneisel, C. (2003b). Electrical resistivity tomography as a tool for geomorphological investigations – some case studies. *Zeitschrift für Geomorphologie Supplement*, **132**, 37–49.

Kneisel, C. (2004). New insights into mountain permafrost occurrence and characteristics in glacier forefields at high altitude through the application of 2D resistivity imaging. *Permafrost and Periglacial Processes*, **15**, 221–227.

Lambiel, C. (2006). *Le pergélisol dans les terrains sédimentaires à forte déclivité: distribution, régime thermique et instabilités.* Institut de Géographie, Université de Lausanne, Travaux et Recherches **33**, 260 pp.

Lambiel, C., Reynard, E., Cheseaux, G. and Lugon, R. (2004). Distribution du pergélisol dans un versant instable, le cas de Tsarmine (Arolla, Evolène, VS), *Bulletin de la Murithienne, Societé Valaisanne des Sciences Naturelles*, **122**, 89–102.

Lugon, R., Delaloye, R., Serrano, E., Reynard, E., Lambiel, C. and González-Trueba, J. J. (2004). Permafrost and Little Ice Age glaciers relationships in the Posets Massif, Central Pyrenees, Spain. *Permafrost and Periglacial Processes*, **15**, 207–220.

Marescot, L., Loke, M. H., Chapellier, D., Delaloye, R., Lambiel, C. and Reynard, E. (2003). Assessing reliability of 2D resistivity imaging in permafrost and rock glacier studies using the depth of investigation index method. *Near Surface Geophysics*, **1**, 57–67.

Reynard, E., Lambiel, C., Delaloye, R., Devaud, G., Baron, L., Chapellier, D., Marescot, L. and Monnet, R. (2003). Glacier/permafrost relationships in forefields of small glaciers (Swiss Alps). *Proceedings of the 8th International Conference on Permafrost*, Zurich, Switzerland, 947–952.

Vonder Mühll, D. (1993). *Geophysikalische Untersuchungen im Permafrost des Oberengadins.* Mitteilungen der Versuchsanstalt für Wasserbau, Hydrologie und Glaziologie, **122**, 222pp.

6

ERT imaging for frozen ground detection

M. Ishikawa

6.1 Introduction

Measurements of direct current (DC) resistivity are useful for detecting frozen ground, because the resistivity values significantly increase in accordance with water's phase change from liquid to solid. Conventional vertical electrical soundings (VES) involve the placement of four-electrode arrays and have been the most commonly used geophysical technique for identifying the occurrence of frozen ground (see Chapter 5). Problems have arisen from the necessary assumption of a horizontally bedded subsurface structure for standard VES interpretation. This situation is further complicated by concave and convex surfaces in topography and significant lateral variation in subsurface water conditions even over small areas, especially in mid-latitude mountainous areas and discontinuous permafrost zones. Therefore, interpretation of data sets obtained by VES alone often remains ambiguous.

As described in detail in Chapter 1, ERT is the hybrid of vertical and horizontal resistivity soundings in which a large number of four-electrode combinations (so-called quadripoles) are measured. This technique two-dimensionally delineates subsurface electrical structures and has recently been recognised as an effective method for investigating frozen ground under complicated topography and/or subsurface structures. In the following case studies ERT was applied in discontinuous and sporadic permafrost areas in order to delineate and characterise frozen ground in the Asian mountain permafrost regions of the Daisetsu Mountains, northern Japan, the Kanchenjunga Himal, eastern Nepal, and the Shijir valley of northeastern Mongolia. On the basis of these research experiences, this note describes practical issues for applying ERT imaging in mountain permafrost regions.

Applied Geophysics in Periglacial Environments, eds. C. Hauck and C. Kneisel. Published by Cambridge University Press. © Cambridge University Press 2008.

6.2 Data acquisition and quality control

Data quality is mostly dependent on the contact resistances between electrode and ground. High contact resistances (i.e. poor conductivity) cause problems such as an insignificant electric current and unstable voltage values, resulting in noisy data sets and making subsequent inversion results uncertain. Experience has shown that the use of steel sticks as electrodes put into the unfrozen fine material generally provides good electrical contact in most cases. On coarse blocky terrain such as rock glaciers, sponges wrapped in steel nets are used instead. Due to the poor conductivity between ground and steel net, however, a high electrical resistance between electrodes and the ground surface is usually encountered. Saturating the sponge electrode with salt water and installing several extra electrodes in parallel might be effective.

Batteries should be recharged as often as possible, because ERT consumes high electric power that is usually supplied by external batteries. However, remoteness of the research sites sometimes does not allow this and the cold climate considerably weakens the batteries. Portable solar panels can be used to recharge the batteries under these conditions, which led to a slight improvement at the Himalayan field site.

Choices of unit electrode spacing and electrode layout determine resolution and maximum penetration depth of the measurements; the correct choice depending on the specific target of the survey. Increasing unit electrode spacing increases the maximum depth for surveys but decreases the resolution. At the sites in the Himalaya and Daisetsu Mountains, where the aim was to delineate the extent and thickness of permafrost and buried ice bodies, we used a unit electrode spacing of 5 m. If the targets are restricted to the active layer and upper part of the permafrost layer, more precise measurements, for example with unit electrode spacing of 1 m, are needed.

Experience has shown that the Wenner–Schlumberger electrode array (see Section 1.2.1), which is moderately sensitive to both horizontal and vertical structure, is the most practical for mountain permafrost sites, allowing acceptable amounts of electric current to be injected. Using a total of 30 electrodes, this array produces more than 400 datum points per survey line. Accordingly, some unreliable and/or noisy data can be neglected with less influence on the final inversion results.

For surface conditions with high electrical resistance, such as completely frozen active layers, the Wenner electrode array might be the only choice, as it is the most robust array against noise. However, this array is less sensitive for detecting vertical subsurface structures. The Dipole–dipole array (see Section 1.2.1), on the other hand, is sensitive to vertical structures, but was found to be inappropriate in the remote mountain sites as it consumes considerably more battery power.

To conduct large numbers of resistivity measurements effectively, as required for ERT, data acquisition should be controlled automatically (Figure 6.1).

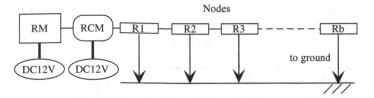

Figure 6.1. View and schematic diagram of the configuration for ERT imaging. RM: resistivity meter (SYSCAL Jr), RCM: remote-control multiplexer. A takeout cable includes 10 nodes in series, through which four electrodes are selected and activated. The system used three takeout cables in total.

Combined usage of multi-core cables, a remote-control multiplexer and a stacking-type digital resistivity meter is preferable. This allows quick and accurate switching of the respective quadripoles. In addition, the sequences of switching active electrodes should be pre-programmed, as is the case for most commercially available multi-electrode systems.

For later analysis noisy and unreliable data must be neglected. For this, data were stacked in accordance with the noise level expressed as a standard deviation of apparent resistivity values. Data with more than 5% standard deviation are categorised as noisy data and normalised by more than 10 times stacking. The noise levels, nevertheless, were not reduced at some datum points, especially where electrodes showed high contact resistances with the ground. All outputs, including the apparent resistivity values, the injected current and the measured voltage, were stored into the resistivity meter along with the converged value of standard deviation. Apparent resistivity values with high standard deviation or insignificant current were excluded from the final analysis data set. In addition, abnormally high or low resistivity values were also neglected (see Section 1.5.1).

6.3 Case studies

The sites of the three case studies are located in the Daisetsu Mountains, northern Japan, the Kanchenjunga Himal, eastern Nepal, and northeast Mongolia, all of which are in the Asian discontinuous permafrost region (Figure 6.2). The geographical settings of each site (i.e. climate, vegetation and topography) are presented in Ishikawa and Hirakawa (2000), Ishikawa *et al.* (2001) and Ishikawa *et al.* (2005), respectively.

6.3.1 Delineating the extent of frozen ground by small-scale surveys: Daisetsu Mountains

The first example (Figure 6.3 (Plate 5)) delineates the extent of mountain permafrost in the Daisetsu Mountains, where permafrost is known to exist underneath the

Table 6.1. *Measurement set-up used for the three study sites shown in Figure 6.2*

		Topography	Unit electrode spacing (m)	Other measurements	Battery recharging
A	Daisetsu Mountains	Mountain ridge	5	SGT[a] profile along survey line	None
B	Kanchen-junga Himal	Rock glacier	5	DC resistivity survey at reference site	Solar panel
C	Shijir Valley	Forested stope	1	Recording soil temperature and moisture	AC

[a] Shallow ground temperature

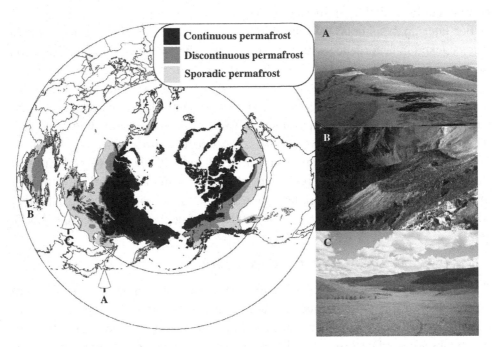

Figure 6.2.   Left: Locations of study sites plotted on the circum-arctic map of permafrost and ground ice conditions (Brown *et al.* 1997). A: Daisetsu Mountains of northern Japan, B: Kanchenjunga Himal of eastern Nepal, C: Shijir Valley of northeastern Mongolia. Right: Views of each location. A: Winter view of the summit areas of the Daisetsu Mountains, B: Rock glacier of Kanchenjunga Himal, C: General landscape of northeastern Mongolia, in which dense forests are restricted on the north-facing slopes where permafrost occurs.

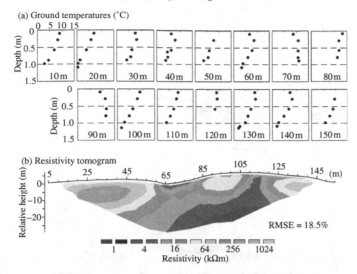

Figure 6.3. (a) Profiles of shallow ground temperature (°C) and (b) inverted DC resistivity (kΩ m) along the same survey line (originally presented in Ishikawa 2003). For colour version see Plate 5.

less snow-covered summit areas (Fukuda and Sone 1992; Figure 6.2A). Along the ERT survey line, shallow ground temperature (SGT) profiles were measured by concurrently boring holes several centimetres in diameter by stick. Even though this method only gives 'rule of thumb' values, it greatly reduced possible ambiguities of permafrost occurrences suggested by the ERT results alone.

The measured SGT decreases with depth for horizontal stations between 10 and 60 m, and between 100 and 150 m (Figure 6.3). SGT measurements at the 120 m station were hampered by shallow bedrock, which made it very difficult to penetrate the ground surface with the hand corer. Ground temperature profiles were equi-thermal between 70 and 90 m. These SGT regimes agreed well with the DC resistivity distribution. High-resistivity zones (>64 kΩ m) were identified between 10 and 60 m, 85 and 105 m and between 125 and 140 m, as probably being frozen. Around 120 m ERT imaging did not suggest the presence of frozen material, as shown by the relatively low resistivity values (4–16 kΩ m). This is probably due to the occurrence of near-surface bedrock, in which resistivity is not as sensitive to freezing and thawing because of small water content.

6.3.2 Delineating buried ice bodies: rock glaciers of the Himalayas

It is generally known that temperate glacier ice shows much higher resistivity than ice-cemented sand-gravels and boulders (e.g. Haeberli and Vonder Mühll 1996). This fact could be used for identifying and delineating glacier-originated

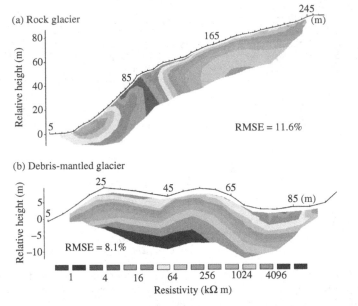

Figure 6.4.   Inverted DC resistivity tomogram of (a) a rock glacier and (b) the
debris-mantled part of the Kanchenjunga Glacier. The unit electrode spacing is
5 m for both measurements (originally presented in Ishikawa *et al.* 2001). For
colour version see Plate 6.

buried ice bodies. The second case study demonstrates this application in the
Kanchenjunga Himal, eastern Nepal, where rock glacier and permafrost co-exist
(Ishikawa and Watanabe 2002; B on Figure 6.2).

Figure 6.4 (Plate 6) shows the inverted resistivity tomogram of the rock glacier
and the debris-mantled parts of the Kanchenjunga Glacier. The inverted resistivity
tomogram demonstrates the existence of extremely high-resistivity ($>1000\,k\Omega\,m$)
zones below the surface of the rock glacier, (Figure 6.4a). Similar order high-
resistivity values were obtained in the debris-mantled glacier (Figure 6.4b). These
facts indicate that a glacier-derived ice body exists within the rock glacier. A low-
resistivity zone below the front slope of the rock glacier probably does not indicate
the ice body, but probably corresponds to unfrozen boulders of the frontal apron.

6.3.3 Monitoring water conditions: Mongolia

Time-lapse ERT measurements using fixed electrodes are known to be a powerful
tool to observe groundwater movements (e.g. Daily *et al.* 1992, Barker and
Moore 1998) and monitor the seasonal changes of unfrozen water content in
mountain permafrost (Hauck 2002, see also Section 1.5.5). The third case study
was obtained in northeastern Mongolia, which is located within the Eurasian

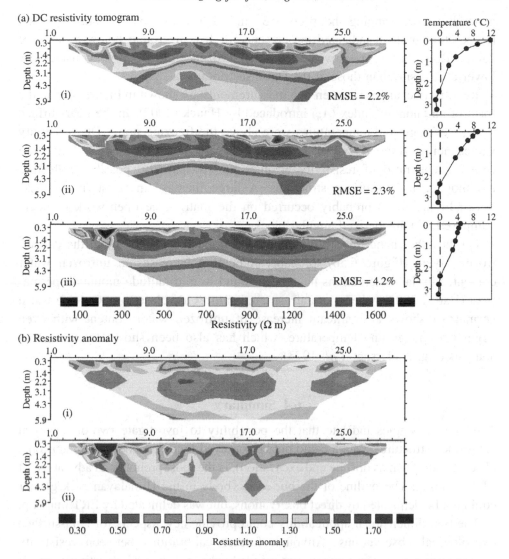

Figure 6.5.   Inversion results of ERT imaging at the permafrost-underlying talus slope, northeast Mongolia. (a) DC resistivity (Ω m) tomogram on (i) 1 July, (ii) 24 August and (iii) 20 September 2003. Daily means of ground temperatures on the respective days are also shown. Unit electrode spacing was 1 m and total number of electrodes was 30. (b) Plots of anomaly index for two consecutive resistivity measurements, showing resistivity changes (i) between 1 July and 24 August, and (ii) between 24 August and 20 September 2003. For colour version see Plate 7.

discontinuous permafrost zone. Permafrost was found only on north-facing forested slopes (Ishikawa *et al.* 2005; C on Figure 6.2).

Figure 6.5a (Plate 7) shows the time series of ERT inversion results in the period of seasonal ground thawing. The calculated resistivity values were generally low, ranging between 0.3 and 1.6 kΩ m, even though ground temperature profiles indicated the presence of frozen ground below a depth of 2.4 m. Remarkably, resistivity values of frozen ground were considerably lower at this site than those at the Daisetsu Mountains.

Resistivity changes between two measurements are shown in Figure 6.5b in the form of an anomaly index (A_n) introduced by Hauck (2002). In the near-surface layer, where ground temperatures were above the freezing point, DC resistivity values increased (i.e. $A_n > 1.0$). This is probably due to a decrease in soil moisture content. The ratio of resistivity changes varied spatially, possibly reflecting inhomogeneous subsurface structures. A large increase in resistivity (i.e. $A_n > 1.5$; Figure 6.5b) probably occurred on the matrix-free open-work boulders composed of talus-derived angular blocks.

On the other hand, resistivity values decreased (i.e. $A_n < 1.0$) in the deeper, frozen layers (Figure 6.5b), probably indicating increases in unfrozen water content. Similar observations have been found in mid-latitude mountain permafrost (Hauck 2002), where permafrost temperatures are close to 0 °C. Such warm permafrost shows a significant increase of unfrozen water content with even slight rises in ground temperature, which has also been shown by laboratory analysis (e.g. Anderson *et al.* 1973).

6.4 Summary

The three examples indicate that the possibility to investigate two-dimensional subsurface structures, which could not be surveyed by direct and one-dimensional observations (such as drilling and excavation), is the most pronounced advantage of ERT imaging. The outline of the buried ice body in the Himalayan rock glacier could not be delineated by direct observations, but was delineated by ERT imaging.

The two-dimensional applicability could provide new aspects for land-surface hydrological observations. Although further calibrations between resistivity values and soil water contents are needed, changes in DC resistivity reflect changes in liquid water content in the active layer and the upper permafrost, both temporally and spatially (Figure 6.5). These characteristics are a major advantage compared to conventional methods for monitoring soil moisture. For example, time-domain reflectometry (TDR) methods measure soil water contents only in the vicinity of the sensors and are not suitable in areas with large inhomogeneities.

The resistivity values of frozen ground were shown to vary significantly between the different sites. Resistivity values of frozen ground at the Mongolian site were two or three orders of magnitude lower than those in the Daisetsu Mountains. Accordingly, mapping of frozen ground by ERT requires supplemental investigations such as SGT measurements or other reference data to reduce the ambiguity in interpretation.

REFERENCES

Anderson, D. M., Tice, A. R. and McKim, H. L. (1973). The unfrozen water and apparent specific heat capacity of frozen soils. *Proceedings of the 2nd International Conference on Permafrost*, Yakutsk, Siberia, 289–295.

Barker, R. and Moore, J. (1998). The application of time-lapse electrical tomography in groundwater studies. *The Leading Edge*, **17**, 1454–1458.

Brown, J., Ferrians, O. J. J., Heginbottom, J. A. and Melnikov, E. S. (1997). *International Permafrost Association circum-Arctic map of permafrost and ground ice conditions, scale 1:10,000,000*. U.S. Geological Survey, Washington, DC.

Daily, W., Ramirez, A., LaBrecque, D. and Nitao, J. (1992). Electrical resistivity tomography of vadose water movement. *Water Resources Research*, **28**(5), 1429–1442.

Fukuda, M. and Sone, T. (1992). Some characteristics of alpine permafrost, Mt. Daisetsu, central Hokkaido, Northern Japan. *Geografiska Annaler*, **74A**, 159–167.

Haeberli, W. and Vonder Mühll, D. (1996). On the characteristics and possible origins of ice in rock glacier permafrost. *Zeitschrift für Geomorphologie, Supplement*, **104**, 43–57.

Hauck, C. (2002). Frozen ground monitoring using DC resistivity tomography. *Geophysical Research Letters*, **29**(21), 2016, doi:10.1029/2002GL014995.

Ishikawa, M. (2003). Spatial mountain permafrost modelling in the Daisetsu Mountains, northern Japan. *Proceedings of the 8th International Conference on Permafrost*, Zürich, Switzerland, 473–478.

Ishikawa, M. and Hirakawa, K. (2000). Mountain permafrost distribution based on BTS measurements and DC resistivity soundings in Daisetsu Mountains, Hokkaido, Japan. *Permafrost and Periglacial Processes*, **11**, 109–123.

Ishikawa, M. and Watanabe T. (2002). Inventory of rock glaciers along the Ghunsa Valley, Kanchanjunga Himal, Eastern Nepal. In International Permafrost Association Standing Committee on Data Information and Communication (comp.). 2003. *Circumpolar Active-Layer Permafrost System, Version 2.0*, eds. Parsons, M. and Zhang, T., Boulder, National Snow and Ice Data Center/World Data Center for Glaciology. CD-ROM.

Ishikawa, M., Watanabe, T. and Nakamura, N. (2001). Genetic differences of rock glacier and the discontinuous mountain permafrost zone in Kanchanjunga Himal, eastern Nepal. *Permafrost and Periglacial Processes*, **12**, 243–253.

Ishikawa, M., Sharkhuu, N., Zhang, Y., Kadota, T and Ohata, T. (2005). Ground thermal and moisture conditions at the southern boundary of discontinuous permafrost, Mongolia. *Permafrost and Periglacial Processes*, **16**, 209–216.

7

Electrical resistivity values of frozen soil from VES and TEM field observations and laboratory experiments

K. Harada

7.1 Introduction

Geophysical approaches have been widely utilised to provide information on permafrost properties or distribution (e.g. Scott *et al.* 1990). The application of geophysical methods to permafrost regions is based on changes of the physical properties of earth materials associated with the freezing of incorporated water. Among them, electrical resistivity values increase greatly when soil water freezes, and electrical sounding methods continue to be used in a number of permafrost studies, as demonstrated in Chapters 1 and 2 as well as in Case Studies 5–10.

It is well known that the electrical resistivity value of soil depends on soil type, temperature, water content, porosity and salinity. To be able to interpret results from resistivity surveys in periglacial environments, it is important to analyse the characteristics of electrical resistivity of frozen soil. Field observations using geophysical methods have been carried out in order to detect permafrost structure and evaluate its applicability for permafrost mapping since 1992. This chapter introduces some observational results from permafrost areas, including resistivity values and their relation to permafrost. In addition, results from corresponding laboratory experiments are shown.

7.2 Methods

The applied geophysical methods are vertical electrical soundings (VES, see Chapter 1) and transient electromagnetic (TEM, see Chapter 2) soundings. A conventional resistivity meter, McOHM model 2115 by OYO Co. Ltd., was used

Applied Geophysics in Periglacial Environments, eds. C. Hauck and C. Kneisel. Published by Cambridge University Press. © Cambridge University Press 2008.

for the VES. The transient data of TEM surveys were recorded using a PROTEM 47 TEM system by Geonics Ltd. with a receiver coil having an effective area of $31.4\,\text{m}^2$.

In order to analyse the physical properties of soil samples from field observation sites, electrical resistivity values were measured in the laboratory under various temperature and water conditions. In this study, four electrodes were arranged in each soil sample within a cylindrical container made of vinyl chloride with 5 cm diameter and 15 cm height. The flat disc current electrodes were located at both ends of the container; the ring potential copper electrodes of 0.5 cm width were located 5 cm from both ends. The same conventional resistivity meter (OYO Co. Ltd., McOHM model 2115) was used as in the field measurements.

7.3 Results

The resistivity profiles obtained from the VES and TEM soundings are shown in Figures 7.1 and 7.2, and values are summarised in Table 7.1. The laboratory results of the soil samples obtained from the field sites are shown in Figure 7.3.

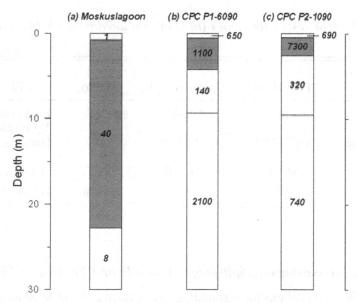

Figure 7.1. Resistivity structures estimated by vertical electrical soundings in (a) Moskuslagoon, Spitsbergen and (b), (c) Caribou-Poker Creeks (CPC), Alaska (Harada and Yoshikawa 1996, Harada *et al.* 2000). Values in the columns show resistivity in $\Omega\,\text{m}$. Shaded parts represent the frozen layer.

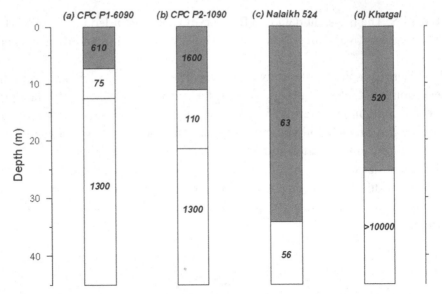

Figure 7.2.   Resistivity profiles estimated by TEM surveys in (a), (b) Caribou-Poker Creeks (CPC), Alaska, (c) Nalaikh, Mongolia, and (d) Khatgal, Mongolia (Harada *et al.* 2000, Harada 2001, Harada *et al.* 2003). Values in the column show resistivity in Ωm. Shaded parts represent the frozen layer.

Table 7.1. *Typical resistivity values (in Ωm) obtained at the field sites of this case study*

Observational site	Permafrost	Unfrozen layer	Method	Reference
Spitsbergen	40	1.1–7.5	DC	Harada and Yoshikawa (1996)
Caribou-Poker Creeks	1100–7300	140–690	DC	Harada *et al.* (2000)
	500–1700	40–230	TEM	Harada *et al.* (2000)
Nalaikh	63	56	TEM	Harada *et al.* (2003)

7.3.1 Moskuslagoon, Spitsbergen (Harada and Yoshikawa 1996)

In Spitsbergen, vertical electrical soundings were carried out at Moskuslagoon near Longyearbyen in 1992 (78° 20′ N, 15° 33′ E). Permafrost is distributed continuously in this area. The mean annual air temperature at Longyearbyen is −4.8 °C, averaged over the period 1961–1990 (Aune 1993). The mean annual ground-surface

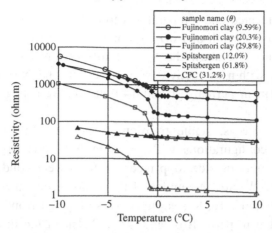

Figure 7.3. Electrical resistivity of soil samples measured in the laboratory for different temperatures and volumetric water contents θ (Harada and Yoshikawa 1996, Harada and Fukuda 2000, Harada *et al.* 2000). Spitsbergen refers to the soil sample obtained from the field site described in Section 7.3.1 and CPC refers to Caribou-Poker Creeks, Alaska (Section 7.3.2).

temperature was $-5.7\,^{\circ}\mathrm{C}$ between 1993 and 1994. Surface material is silty clay with high water content. The depth of the active layer was 0.8 m in September 1992. No vegetation covered the surface, because this area had been temporarily immersed in seawater.

Estimated resistivity values (Figure 7.1a) ranged between 1 (unfrozen) and $40\,\Omega\,\mathrm{m}$ (permafrost), which is an extremely low value for frozen ground (see Chapter 1 Table 1.1 and Table A2). The laboratory results shown in Figure 7.3 confirm that the low values obtained at the field site were caused by high salinity. The salinity and resistivity values of soil water obtained from the study site at $25\,^{\circ}\mathrm{C}$ were 17% and $0.33\,\Omega\,\mathrm{m}$, respectively. This high salinity also caused the freezing point depression (Figure 7.3). The estimated permafrost depth was 22.8 m. Using the obtained permafrost depth, a numerical calculation of the permafrost age was conducted. The calculated age appears to be reasonable compared to the age derived from radiocarbon dating of adjacent sediments (Harada and Yoshikawa 1996).

7.3.2 Caribou-Poker Creeks, Alaska (Harada et al. 2000)

The Caribou-Poker Creeks are located in central Alaska ($65^{\circ}\ 10'\,\mathrm{N}$, $147^{\circ}\ 30'\,\mathrm{W}$). Mean annual air temperature is about $-4.0\,^{\circ}\mathrm{C}$ and permafrost is discontinuously distributed. Field observations were carried out in 1998. Vertical electrical soundings and TEM surveys were conducted at the same points, named P1-6090 and P2-1090, in the permafrost zone with spruce and moss surface cover. Surface material is clayey silt. Near the measurement points, boreholes for temperature monitoring were located and temperatures to a depth of 8 m were continuously measured indicating a depth of frozen ground greater than 8 m.

Figures 7.1b and 7.1c show the VES results obtained at the two locations. Resistivity values within the active layer were 650 and 690 Ω m, respectively, 1100 and 7300 Ω m in permafrost and 140 and 320 Ω m in the unfrozen layer below.

For the TEM surveys, stable transient responses were recorded only for the outside loop configurations with a single turn loop of 60 m × 60 m (see Chapter 2). Thus, only outside loop measurements were conducted at this site. Figures 7.2a and 7.2b show examples of the TEM results at the same locations as in Figures 7.1b and 7.1c. Resistivity values obtained from the TEM surveys were 500–1700 Ω m in permafrost and 40–230 Ω m in the unfrozen layer. The comparatively small resistivity difference between frozen and unfrozen material may be caused by the clayey silt material which is present near the surface at this site.

Comparing the resistivity profiles obtained by the two methods, the depth of the permafrost base estimated by the VES method was less than 5 m and shallower than that obtained by the TEM method (more than 8 m). The borehole temperature profiles support the accuracy of the resistivity structure obtained by the TEM method. In order to investigate the cause of this difference, inversions of TEM and VES electrical data were generated from two types of modelling. In the first inversion scheme, a synthetic voltage data set with 3% random noise added to the one-dimensional profiles was used for both methods. In the second inversion scheme, a DC electrical data set estimated from a resistivity structure with horizontal variability was used. From the results of these inversions, it was concluded that the TEM method can show the subsurface structure more precisely concerning the estimation of the permafrost thickness. More details can be found in Harada *et al.* (2000).

7.3.3 Nalaikh, Mongolia (Harada et al. 2003)

Nalaikh (47° 45' N, 107° 15' E) is located in the suburbs of Ulaanbaatar, central Mongolia. Permafrost is discontinuously distributed and the permafrost base at the survey site was located at a depth of 36.4 m based on the temperature profiles of a borehole, named 524. TEM surveys were conducted near borehole 524 in 1998 by using a transmitter loop of 60 m × 60 m. Measurements in the central loop configuration were used at this site (see Chapter 2).

The obtained resistivity showed low values, and the vertical variation of resistivity values was small (Figure 7.2c). Again, the resistivity difference between frozen and unfrozen layers was very small. This small difference may have been caused by low water content. The estimated permafrost base ranged from 34 m to 45 m, which was in good agreement with the depth from the ground temperature profile, about 36 m, even though the resistivity difference between frozen and unfrozen layers is small.

7.3.4 Khatgal, Mongolia (Harada 2001)

Khatgal (50°27'N, 100° 10' E) is located in Northern Mongolia in the continuous permafrost zone. Mean annual air temperature is − 4.8 °C (1963–1992). TEM surveys were conducted in 1998 on the west side of Khovsgl Lake near a borehole. The permafrost thickness was extrapolated as 60–70 m based on the borehole temperature profile down to 9.5 m.

The obtained resistivity structure shows resistivity values for the surface frozen layer of about 500 Ω m, and high resistivity values (more than 10 000 Ω m) for the second layer (Figure 7.2d). However, the thickness of the surface layer ranged from 23 to 28 m, which did not coincide with the estimated thickness using the borehole temperature profile (60–70 m). At this site, the second layer had a much higher resistivity value than the first layer, and may represent bedrock, although the depth of bedrock has not been reported. As porosity and water content of bedrock are very small, bedrock resistivity can be very high even in the unfrozen state. Therefore, the resistivity difference between frozen and unfrozen rocks is small, and the permafrost base could not be detected by using TEM or other electrical sounding methods.

7.3.5 Laboratory experiments (Harada et al. 1994, Harada and Fukuda 2000, Harada 2001)

Figure 7.3 shows the experimental results of soil samples obtained from the various field sites. The Spitsbergen sample is classified as silty clay, the Caribou-Poker Creeks (CPC) sample as clayey silt. In Figure 7.3, the results for Fujinomori clay are also shown, which is classified as clayey silt with high unfrozen water content near 0 °C.

These experimental results on fine-grained soils show that the resistivity difference between frozen and unfrozen soils is small due to the presence of unfrozen water, especially in the case of low water content. Based on these experimental results, Harada and Fukuda (2000) developed a resistivity estimation model for frozen soil as a function of temperature. In the case of unsaturated frozen soil, it was assumed that electrical conduction was divided into two parts: (i) soil surface conduction and (ii) conduction through unfrozen water. The model is composed of the elements ice, soil particles and air. The bulk resistivity of soil

ρ can then expressed as

$$\frac{1}{\rho} = \frac{1}{\rho_{\text{sur}}} + \frac{b}{\rho_{\text{u}}} + \frac{c}{\rho_{\text{i}}} + \frac{d}{\rho_{\text{s}}} + \frac{e}{\rho_{\text{a}}}, \qquad (7.1)$$

where ρ_{sur} is the resistivity for soil surface conduction, ρ_{u}, ρ_{i}, ρ_{s} and ρ_{a} are the resistivities of unfrozen water, ice, soil particles and air, respectively. The parameters b, c, d and e are the fractional cross-sectional areas of each element. If soil particles and air are regarded as electrical insulators, Equation (7.1) becomes

$$\frac{1}{\rho} = \frac{1}{\rho_{\text{sur}}} + \frac{b}{\rho_{\text{u}}} + \frac{c}{\rho_{\text{i}}}. \qquad (7.2)$$

As each resistivity value, ρ_{sur}, ρ_{u} and ρ_{i}, can be expressed as a function of temperature, this equation can also be written simply as a function of temperature. Using this estimation model, Harada and Fukuda (2000) showed that the calculated resistivity values are in good agreement with the measured values.

7.4 Summary

From field observations and laboratory experiments, various electrical resistivity values for different permafrost soils were obtained, which range widely from around $40\,\Omega\,\text{m}$ (frozen soils with high salinity), over $1000–7000\,\Omega\,\text{m}$ (frozen clayey silt with low salinity) to more than $10\,000\,\Omega\,\text{m}$ in the case of frozen bedrock. Moreover, a comparison between VES and TEM sounding results showed discrepancies concerning both layer depth and the specific resistivities of each layer. Forward-inverse modelling (see Chapter 1) and ground truth provided by borehole temperatures showed a more precise estimation of layer thicknesses by the TEM method. Permafrost thicknesses estimated from borehole temperature data revealed, furthermore, that estimation of the permafrost base by TEM soundings failed if the contrast between frozen and unfrozen material (e.g. bedrock with low water content) was small (see also Chapter 2). Thus, laboratory and model approaches are required to clarify the resistivity characteristics of frozen ground in order to relate observed resistivity soundings to the permafrost distribution.

REFERENCES

Aune, B. (1993). Temperaturnormaler. *Det Norske Metorologiske Institutt Klima Rapport*, **51**, 2–93.
Harada, K. (2001). *Studies on Detection of Permafrost Structure*. Ph.D. Thesis, Hokkaido University, 163pp.

Harada, K. and Fukuda, M. (2000). Characteristics of the electrical resistivity of frozen soils. *Seppyo*, **62**, 15–22 (in Japanese with English abstract).

Harada, K. and Yoshikawa, K. (1996). Permafrost age and thickness near Adventfjorden, Spitsbergen. *Polar Geography*, **20**, 267–281.

Harada, K., Ishizaki, T. and Fukuda, M. (1994). Measurement of electrical resistivity of frozen soils. *Proceedings of the 7th International Symposium on Ground Freezing*, Nancy, France, 153–156.

Harada, K., Wada, K. and Fukuda, M. (2000). Permafrost mapping by transient electromagnetic method. *Permafrost and Periglacial Processes*, **11**, 71–84.

Harada, K., Wada, K. and Fukuda, M. (2003). Detection of permafrost structure by transient electromagnetic method in Mongolia. *Proceedings of the 8th International Conference on Permafrost*, Zürich, Switzerland, Extended Abstracts Reporting Current Research and New Information, 53–54.

Scott, W., Sellmann, P. and Hunter, J. (1990). Geophysics in the study of permafrost. In *Geotechnical and Environmental Geophysics*, ed. Ward, S., Society of Exploration Geophysics, Tulsa, pp. 355–384.

8

Results of geophysical surveys on Kasprowy Wierch, the Tatra Mountains, Poland

W. Dobinski, B. Zogala, K. Wzietek and L. Litwin

8.1 Introduction

Geophysical surveys are more and more commonly used for the investigation of both permafrost features and their changes in high mountain environments. Before the rebuilding of the cable railway on Kasprowy Wierch, Tatra Mountains, Poland, geophysical techniques were applied to determine the ground conditions of the planned construction site. The survey results were used to prepare a geotechnical expert report for the reconstruction of the upper section of the cable railway. This contribution aims to show the possibilities, limitations and ways of interpreting the geophysical measurements obtained in a mountainous periglacial environment within the zone of discontinuous permafrost.

8.2 Field site

Kasprowy Wierch is a peak in the main ridge of the Tatra Mountains (Figure 8.1), 1986 m a.s.l. It consists of granodiorite and pegmatites, which form a tectonic cap-rock on the summit with a thickness of a few hundred metres (Bac-Moszaszwili and Gąsienica-Szostak 1990). Faults and fractures can be seen in the dome. Although the area of the Tatra Mountains was glaciated several times (Gadek 1998), the peak was transformed by periglacial processes only (Klimaszewski 1988), which led to the creation of block fields on the dome. The thickness of the weathered material reaches 3–4 m (Gryczmanski *et al.* 2004). In contrast to the peak, all adjacent valleys were glaciated. Distinct glacial undercutting rock faces are visible on the northern part of the peak (Figure 8.2). Kasprowy Wierch is the only mountain in the Polish Tatra Mountains so extensively used by people. On the summit, an upper section of the cable railway as well as a meteorological observatory are located,

Applied Geophysics in Periglacial Environments, eds. C. Hauck and C. Kneisel. Published by Cambridge University Press. © Cambridge University Press 2008.

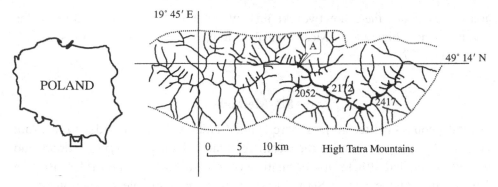

A Kasprowy Wierch summit, 1986 m a.s.l.

⌒⌒⌒   Main ridge of the High Tatra Mountains

·······   Border of the Tatra Massif

Figure 8.1.   Location of the research area.

Figure 8.2.   The field site of Kasprowy Wierch summit in a north-oriented air photograph with locations of geophysical surveys indicated: continuous line for DC resistivity tomography, dashed lines for shallow electromagnetic survey, and numbered points for VES. A: upper ski-lift station on the Goryczkowy cirque, B: meteorological station at the top of Kasprowy Wierch, C: upper cable-car station, D: upper ski-lift station on the Gasienicowy cirque.

and on its slopes there are two ski lifts, whose top stations are visible on the satellite image (Figure 8.2). Mean annual air temperature on the summit is –0.8 °C (Dobinski 1997).

8.3 Methods

Several geophysical techniques were applied and combined in order to measure various physical properties of the weathered material and underlying bedrock and to obtain detailed structural information on the subsurface. The BTS (Bottom Temperature of the winter Snow cover) method may enable prediction of the occurrence of permafrost if the thickness of the active layer is no more than 6 m (Haeberli 1973). The EM-31 electromagnetic method can penetrate to similar depths (Hauck *et al.* 2001; see Chapter 2). On the other hand, VES and ERT (Chapter 1) surveys may reach penetration depths from several metres to more than 30 m (Vonder Mühll *et al.* 2001). These methods have been successfully used for many years in both alpine and arctic periglacial environments. The terrain model was generated on the basis of a high-resolution multispectral satellite image from the IKONOS satellite, taken on 28 August 2004, and a digital terrain model based on topographic maps (1:10 000), kindly made available by the Institute of Spatial and Cadastral Systems and the management of the Tatra National Park.

8.4 Measurements

8.4.1 BTS

Twenty-eight measurements were performed in the summit area: on the north and south sides of the Kasprowy Wierch–Beskid ridge, in the upper section of Gasienicowy cirque, on the summit dome, and on the northwest slope of Kasprowy Wierch (Figure 8.3). Measurements were carried out at the beginning of March and were restricted to locations where the dry snow cover thickness was more than 80 cm from the beginning of January. Spatial distance between BTS measurements was about 20–30 m.

On the southern side, observed BTS values were higher than −2 °C, whereas they were much lower, even reaching −5 °C and below, at the northern side at similar or even lower altitudes. The spatial pattern of BTS on the summit dome of Kasprowy Wierch was similar (but with smaller amplitude) to the one in the ridge area. BTS ranged from −2.7 °C in south exposed places, −3 °C on the summit and −3.2 °C in places of slightly northwest exposure. This difference might be significant for permafrost occurrence; however, the accuracy of the PT100

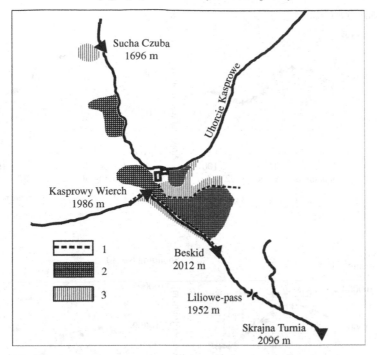

Figure 8.3. Results of the BTS measurements drawn on a north-oriented location map. 1: places with BTS lower than –3 °C, 2: places with BTS temperatures between –2 and –3 °C, 3: places with temperatures higher than –2 °C.

probes is only around 0.1 °C. In places with favourable conditions, such as small depressions with relatively large amounts of quarternary sediments and northern or northwestern exposure, BTS values were as low as -3.8 °C to -4.3 °C (Dobinski 2004a).

8.4.2 VES

Vertical electrical soundings in a Schlumberger symmetrical configuration were carried out on the east-exposed slope and on the ridge near the Kasprowy Wierch summit (Figure 8.2). The results of all performed soundings were rather diverse, so that a uniform (similar) interpretation for all sounding results cannot be given. The most typical sounding curves for mountain permafrost are shown in VESs 38 and 39 (Figure 8.4), where beneath a 1 m thick blocky cover (with resistivity around 10 kΩ m) a 1.5–3 m thick layer is found, which is characterised by high resistivity values of 60–70 kΩ m (cf. Chapters 5 and 9). However, as the weathered layer reaches 3–4 m thickness, the high-resistivity layer seems to be confined to the weathered, blocky material, which could cause high resistivities even in the absence of frozen material (cf. Chapter 1). Two

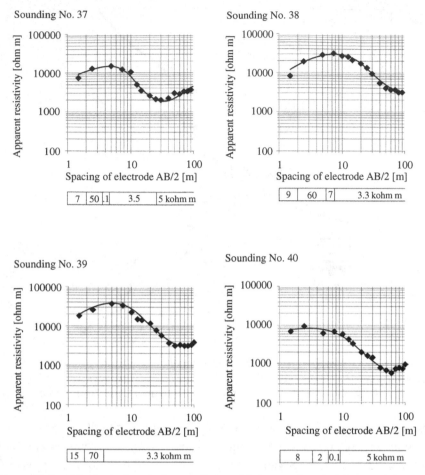

Figure 8.4.   Vertical electrical soundings with inverted resistivity model.

high-resistivity anomalies can be observed in VES 37: the first one is a layer with a specific resistivity of 50 kΩ m and a thickness of 1.5 m located beneath 1 m of covering sediments (feature similar to VESs 38 and 39), the second one (about 5 kΩ m) occurs deeper than 11 m; they are separated by a low-resistivity layer of 0.1–3.5 kΩ m. The resistivity in this lowermost layer seems to increase to even higher values at greater depth. A similar situation in VES 40 is clearly visible. Resistivity increases again at a depth of 9 m to values of more than 5 kΩ m (Figure 8.4).

8.4.3 ERT

ABEM DC resistivity equipment was used for ERT data acquisition. A 200 m long survey line was laid out with an electrode spacing of 2.5 m (Figure 8.5 (Plate 8)).

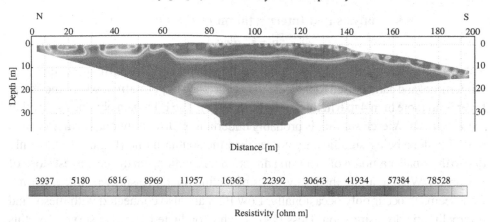

Figure 8.5. DC resistivity tomography profile. For colour version see Plate 8.

Measurements were conducted using the Wenner array and a current intensity ranging from 20 to 200 mA, an acquisition delay of 0.3 s, an acquisition time of 1.0 s and minimum/maximum number of measuring cycles of 2 and 4, respectively, with a standard deviation criterion between subsequent measuring cycles of maximal 5% (cf. Chapter 1). Most data points showed standard deviations of a small percentage, with an occasional increase to around 10%. Greatest diversity in the values was observed for the largest electrode spacings. Field data were inverted using RES2DINV software with both the least-squares smoothness as well as the robust model constraint as inversion methods (Claerbout and Muir 1973, deGroot-Hedlin and Constable 1990). The robust method was chosen because it allows better distinction of subsurface boundaries. Inversions were performed using five iterations resulting in a specific resistivity cross-section with a root-mean-square (RMS) error of 3.7% (Figure 8.5). The Oasis Montaj graphic program was applied to generate the cross-section.

8.4.4 Electromagnetic surveys

The shallow electromagnetic measurements were made using an EM31-MK2 conductivity meter (Geonics). Three parallel survey lines of 105 m length, spaced 20 m apart, were measured (Figure 8.2). The profile method was deployed, the data were recorded automatically every 0.5 s, which gave data points every 0.2 to 0.4 m on each line. The vertical coplanar configuration of the measuring coils (VCP, see Chapter 2, Figure 2.3b) was used with a penetration depth of 6 m. The Oasis Montaj graphic program was applied to present the results as horizontal map of bulk apparent electrical conductivity.

8.5 Analysis and interpretation of the measurements

8.5.1 BTS

The BTS measurements allow mapping of the ground surface thermal conditions in wintertime, and can be used to indicate the occurrence of permafrost where the active layer is no more than 4 to 6 m thick (Haeberli 1973). The BTS data obtained show that the Kasprowy Wierch summit is probably underlain by discontinuous permafrost, the northern slope being significantly cooler than the southern one (Figure 8.3), mainly due to the smaller amount of incoming direct solar radiation. On the northwest slope of Kasprowy Wierch, along the rocky walls of Dolina Sucha Kasprowa valley, permafrost seems to occur only occasionally. Low BTS are also connected with meso- and microclimatic diversification. Hess (1974), who conducted climatic surveys on this slope of Kasprowy Wierch, argues that in the ground depressions the mean annual air temperature can be below zero (−0.4 °C) even at an altitude of 1700 m. The BTS measurements indicate that permafrost can also be present in such geomorphic features. On Kasprowy Wierch the thickness of continuous winter snow cover reaches 2.5 m in places and air circulation within the slope deposits is suppressed due to the presence of a ski run with its particularly dense snow cover. These snow conditions do not fully support the existence of permafrost, but early winter freezing of the ground, and relatively low BTS values do not allow this hypothesis to be excluded.

8.5.2 Electromagnetic surveys

Electrical conductivities measured on Kasprowy Wierch are characterised by low values ranging from about 0.2 to 1.3 mS/m (Figure 8.6 (Plate 9)). Such low values

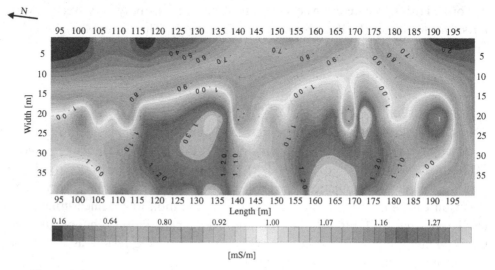

Figure 8.6.   Map of the EM-31 conductivity results. For colour version see Plate 9.

and their slight variation are typical of high mountains where permafrost occurs (Hauck *et al.* 2001). The lowest values of electrical conductivity were recorded in the lower section of the slope. Here granodiorite boulders with numerous air-filled spaces between can be seen on the surface. Their presence results in a decrease in apparent electrical conductivity due to the electrically isolating properties of air. In the upper sections of the surveyed slope (rockslides with rubble), additional regions with significantly lower conductivity values can be found. Places with outcrops of granodiorite showed slightly higher apparent electrical conductivity. Very low conductivity values shown in Figure 8.6 can be the effect of ice accumulation at a depth of 2–3 m within the blockfield (cf. Figure 8.5).

8.5.3 VES

The interpretation of the various vertical electrical soundings was difficult due to the great diversification of the measurements (Figure 8.4). VESs 38 and 39 have characteristics of very shallow mountain permafrost (Haeberli 1985, Vonder Mühll 1993, Kneisel 2003), while sounding 37 reveals the presence of two high-resistivity layers. When compared to the ERT results (Figure 8.5) the high resistivity values in this sounding (50 and 5 kΩ m) might indicate the possibility of the existence of two permafrost layers. Similar findings have been reported from Jotunheimen, Norway, by Etzelmüller *et al.* (1998, 2001).

8.5.4 ERT

The ERT imaging method enabled the detection of the two-dimensional resistivity distribution. Resistivity values range from a few kΩ m to tens of kΩ m. A discontinuous surface layer ranging from around 3 m to 5.5 m depth is visible (Figure 8.5). It is interpreted as a boundary between the weathered material and the massive granodiorite below. As for VES, the layer is characterised by resistivity values of around 15 kΩ m. It is built of loose mixed material, partially humid. Resistivity values like this are characteristic of the superficial debris layer in alpine areas (e.g. Evin and Fabre 1990, Vonder Mühll 1993). A few anomalies of very high resistivity values are delineated above and below this boundary.

The high resistivity values near the surface reach 30 to 90 kΩ m and are characteristic of blockfields with numerous air voids (Barsch and King 1989) as well as permafrost occurrences (Dobinski 1997, Kedzia *et al.* 1998, Vonder Mühll *et al.* 2001, Kneisel and Hauck 2003, Kneisel 2006). The layer has a thickness of 3–4 m and its upper boundary is at a depth of 1–2 m. The horizontal lenticular arrangement of anomalies of highest resistivity can be regarded as permafrost, whereas singular, small high resistivity anomalies near the surface are interpreted as air voids.

The layers below the delineated boundary can be treated as a uniform granodiorite massif with characteristic resistivity of around 4 kΩ m. In its upper section the presence of fractures is possible, including tectonic ones, reaching depths from tens to even hundreds of metres. In this bedrock zone, at a depth of 15–25 m, another high-resistivity anomaly with resistivities of 15–20 kΩ m can be distinguished. One possible interpretation for this high-resistivity layer could be the presence of permafrost in the degradation stage. If measurement or inversion errors can be excluded, a purely lithological interpretation is difficult to adopt, since tectonic discontinuities in high mountains are more likely to be infiltrated by water and would result in lower resistivity values. If a deep permafrost layer exists, it should be significantly older than the possible near-surface layer (Dobinski 2004b).

8.6 Conclusions

Geophysical surveys were conducted by means of several independent methods including BTS, electromagnetic induction (EM31), vertical electrical soundings (VES) and electrical resistivity tomography (ERT). They all tend to indicate the possible occurrence of a thin layer of discontinuous permafrost on Kasprowy Wierch at 2–4 m depths beneath the ground surface.

The methods which allow for deeper penetration of the ground, that is VES and ERT, both indicate the occurrence of a possible high-resistivity anomaly at a depth of around 11 m (VES) or between 15 and 25 m (ERT) below the surface. It may be related to relict permafrost or to a change in the mineral composition of the bedrock.

Although the data obtained may indicate a fossil permafrost occurrence on Kasprowy Wierch, this has to be confirmed by applying other methods, both direct (borehole drilling) or indirect (refraction seismics, GPR).

Acknowledgments

The authors highly appreciate useful improvements made by C. Hauck and C. Kneisel in the earlier version of the manuscript.

REFERENCES

Bac-Moszaszwili M. and Gąsienica-Szostak M. (1990). *Tatry Polskie przewodnik Geologiczny dla Turystów*. Wydawnictwa Geologiczne, Warszawa.
Barsch D. and King L. (1989). Origin and geoelectrical resistivity of rock glaciers in semi-arid subtropical mountains (Andes of Mendoza, Argentina). *Zeitschrift für Geomorphologie, Supplement*, **33**(2), 151–163.

Claerbout, J. F. and Muir, F. (1973). Robust modeling with erratic data. *Geophysics*, **38**, 826–844.

deGroot-Hedlin, C. and Constable, S. (1990). Occam's inversion to generate smooth two-dimensional models from magnetotelluric data. *Geophysics*, **55**, 1613–1624.

Dobinski, W. (1997). Distribution of mountain permafrost in the High Tatras based on freezing and thawing indices. *Biuletyn Peryglacjalny*, **36**, 29–37.

Dobinski, W. (2004a). Granica występowania wieloletniej zmarzliny w Tatrach. *Czasopismo Geograficzne*, **75**(1–2), 123–132.

Dobinski, W. (2004b). Wieloletnia zmarzlina w Tatrach: geneza, cechy, ewolucja. *Przegląd Geograficzny*, **76**, 327–343.

Etzelmüller, B., Berthling, I. and Sollid, J. L. (1998). The distribution of permafrost in southern Norway – a GIS approach. *Proceedings of the 7th International Conference on Permafrost*, Yellowknife, Canada, 251–257.

Etzelmüller, B., Hoelzle, M., Solbjørg Flo Heggem, E., Isaksen, K., Mittaz, C., Vonder Mühll, D., Ødegard, R. S., Haeberli, W. and Sollid, J. L. (2001). Mapping and modelling the occurence and distribution of mountain permafrost. *Norsk Geografisk Tidsskrift*, **55**, 186–194.

Evin, M. and Fabre, D. (1990). The distribution of permafrost in rock glaciers of the southern Alps (France). *Geomorphology*, **3**, 57–71.

Gadek, B. (1998). *Würmskie zlodowacenie Tatr w świetle rekonstrukcji lodowców wybranych dolin na podstawie prawidłowości glacjologicznych*. Wydawnictwo Uniwersytetu Śląskiego, Katowice.

Gryczmanski, M., Dobinski, W. and Sołowski, W. (2004). *Ekspertyza geotechniczna dla potrzeb przebudowy kolei linowej na Kasprowy Wierch (odcinek Myślenickie Turnie – Kasprowy Wierch)*. Instytut Systemów Przestrzennych i Katastralnych, Gliwice.

Haeberli, W. (1973). Die Basis-Temperatur der winterlichen Schneedecke als moglicher indikator für die Verbreitung von permafrost in den Alpen. *Zeitschrift für Gletscherkunde und Glazialgeologie*, **9**(1–2), 221–227.

Haeberli, W. (1985). *Creep of Mountain Permafrost: Internal Structure and Flow of Alpine Rock Glaciers*. Mitteilungen der Versuchsanstalt für Wasserbau, Hydrologie und Glaziologie, 77, 143pp.

Hauck, C., Guglielmin, M., Isaksen, K. and Vonder Mühll, D. (2001). Applicability of frequency-domain and time-domain electromagnetic methods for mountain permafrost studies. *Permafrost and Periglacial Processes*, **12**, 39–52.

Hess, M. (1974). Piętra klimatyczne Tatr. *Czasopismo Geograficzne*, **45**(1), 75–93.

Jahn, A. (1970). *Zagadnienia Strefy Peryglacjalnej*. PWN Warszawa.

Kedzia, S., Moscicki, J. and Wrobel, A. (1998). Studies on the occurence of permafrost in Kozia Valley (the High Tatra Mts.). Wyprawy Geograficzne na Spitsbergen, IV Zjazd Geomorfologów Polskich UMCS, Lublin 3-6 Czerwca 1998, 51–57.

Klimaszewski, M. (1988). *Rzeźba Tatr Polskich*. PWN Warszawa.

Kneisel, C. (2003). Permafrost in recently deglaciated glacier forefields – measurements and observations in the eastern Swiss Alps and northern Sweden. *Zeitschrift für Geomorphologie*, **47**, 289–305.

Kneisel, C. (2006). Assessment of subsurface lithology in mountain environments using 2D resistivity imaging. *Geomorphology*, **80**, 32–44.

Kneisel, C. and Hauck, C. (2003). Multi-method geophysical investigation of a sporadic permafrost occurence. *Zeitschrift für Geomorphologie, Supplement*, **132**, 145–159.

Vonder Mühll, D. (1993). *Geophysikalische Untersuchen im Permafrost des Oberengadins*. Mitteilungen der Versuchsanstalt für Wasserbau, Hydrologie und Glaziologie, 122, 222pp.

Vonder Mühll, D., Hauck, C., Gubler, H., McDonald, R. and Russill, N. (2001). New geophysical methods of investigating the nature and distribution of mountain permafrost with special reference to radiometry techniques. *Permafrost and Periglacial Processes*, **12**(1), 27–38.

9

Reassessment of DC resistivity in rock glaciers by comparing with P-wave velocity: a case study in the Swiss Alps

A. Ikeda

9.1 Introduction

Vertical electrical resistivity soundings (VES) have been extensively used to investigate the internal structure of rock glaciers (e.g. Fisch *et al.* 1977, King *et al.* 1987, Ikeda and Matsuoka 2002). In particular, two-dimensional (2D) electrical resistivity tomography (ERT) has recently revealed the internal structure in greater detail (e.g. Vonder Mühll *et al.* 2000, Ishikawa *et al.* 2001, Ikeda and Matsuoka 2006, see also Chapters 1 and 6). Hauck and Vonder Mühll (2003), however, argued that even such tomographical methods include ambiguity in the model inversion and interpretation of DC resistivity. Although combining different geophysical properties reduces ambiguity in the interpretation of a single geophysical property (Haeberli 1985, Hauck and Vonder Mühll 2003), many studies have relied on DC resistivity methods alone. Thus, interpretation of DC resistivity in rock glaciers and related terrain requires further assessment.

This study compares DC resistivity with P-wave velocity in a number of rock glaciers. DC resistivities and/or P-wave velocities were measured on 26 talus-derived rock glaciers lying near the lower limit of mountain permafrost in the Upper Engadin, Swiss Alps. Both one-dimensional (1D) VES and 2D ERT surveys were performed at seven sites. Special attention was given to the effects of spatial variations in lithological and thermal conditions. The results were compiled for three types of rock glaciers:

(i)   bouldery rock glaciers having an active layer composed of matrix-free boulders,
(ii) pebbly rock glaciers consisting of matrix-supported pebbles and cobbles, and
(iii) (densely) vegetated rock glaciers that mostly represent relict (bouldery) rock glaciers (see Ikeda and Matsuoka 2002, 2006 for the classification).

Applied Geophysics in Periglacial Environments, eds. C. Hauck and C. Kneisel. Published by Cambridge University Press. © Cambridge University Press 2008.

The first two types lack vegetation on their upper surface. Weathering-susceptible rockwalls above pebbly rock glaciers produce much finer debris than resistant rockwalls above bouldery rock glaciers. Although data on geophysical properties have been obtained extensively from bouldery rock glaciers (e.g. Barsch 1996, Haeberli and Vonder Mühll 1996, Vonder Mühll *et al.* 2002), such data from pebbly and vegetated (relict) rock glaciers is lacking. Further details about the studied rock glaciers (locations, abbreviations, photographs, topographical maps/ profiles, etc.) can be found in Ikeda and Matsuoka (2002, 2006) and Ikeda (2006).

9.2 Methods

A lightweight seismograph, McSEIS-3 (manufactured by OYO, Japan), was used for the refraction seismic soundings on 21 sites. A 4 kg sledgehammer was used to produce seismic pulses. P-wave velocities of two or three layers and the depth(s) of the layer boundary (boundaries) were determined using the intercept-time method (Palmer 1986; see also Chapter 3). The reciprocal method (Palmer 1986) was also employed to obtain more accurate P-wave velocities of the second layer by eliminating anomalies caused by irregular ground and refracting surfaces. The latter method was available only when a pair of traveltime curves indicated two-layer structure with the layer boundary at a relatively shallow depth (1–5 m deep in this study). The soundings were carried out from mid July to mid August 2000–3.

One-dimensional VES surveys were performed on 23 sites from late July to early August in 1999 and 2002 with the SYSCAL R1 Plus resistivity meter (Iris Instruments, France). Inversions of the observed apparent resistivity curves to yield matching modelled resistivity curves were calculated with WinSev6 software (W_GeoSoft, Switzerland).

Two-dimensional DC resistivity distributions were obtained by ERT surveys for two bouldery and five pebbly rock glaciers. The electrode configuration followed the Wenner–Schlumberger array, which combines the Wenner and Schlumberger configurations in a joint profile (e.g. Ishikawa *et al.* 2001; see also Chapter 1). The ERT profiles were set parallel to the flow line of the rock glaciers. The surveys were carried out with the SYSCAL Junior resistivity meter (Iris Instruments, France) from mid July to early August in 2001 and 2002. Two-dimensional distributions of modelled DC resistivities were computed with the RES2DINV (ver. 3.4) software (Geotomo Software, Malaysia).

9.3 Field sites with borehole information

The sounded sites include two active rock glaciers where the internal structures were directly observed: Murtèl (M1) bouldery rock glacier (e.g. Haeberli *et al.*

1988, Arenson *et al.* 2002; see also Chapter 1) and the upper lobe of Büz North (BNU) pebbly rock glacier (Ikeda and Matsuoka 2006, Ikeda *et al.* 2008). M1 has an approximately 30 m thick ice-rich layer underlying a 2.5–3 m thick layer of matrix-free boulders and overlying frozen coarse debris down to 50 m deep near the sounding site. The upper half of the ice-rich layer is pure ice, whereas the lower half includes silt, sand and gravels in a highly ice-supersaturated structure. The bottom of the ice-rich layer consists of a 4 m thick layer of frozen sand and silt. Mean annual temperatures within the ice-rich layer are -1.5 to $-2\,°C$. BNU has slightly ice-supersaturated pebbles and cobbles below a 2–2.5 m thick pebbly active layer. The frost table of BNU lay about 1 m deep during the sounding periods and the temperature below the permafrost table was close to $0\,°C$ throughout the year.

9.4 Results

9.4.1 P-wave velocity

Two to three velocity layers were identified for each pair of traveltime curves (Table 9.1, Figure 9.1). A sharp break in P-wave velocity was observed in 17 pairs of traveltime curves. The break results from the large contrast in velocity between the upper layer(s) (330–$730\,\mathrm{m\,s^{-1}}$) and the lower layer (2000–$4400\,\mathrm{m\,s^{-1}}$), which represents the layer boundary between the upper unfrozen debris and lower frozen debris or bedrock. The other four curves lack such a break and have low P-wave velocities (360–$950\,\mathrm{m\,s^{-1}}$) over the whole investigation depth, except for an intermediate velocity ($1600\,\mathrm{m\,s^{-1}}$) below 7–12 m deep in PN3.

Calculated thicknesses of the low-velocity layers are thinner than 5 m on six bouldery and seven pebbly rock glaciers, whereas they are about 10 m or more on five vegetated rock glaciers. In the former cases, the lower layer with high P-wave velocity probably corresponds to frozen debris or ice, because the upper surface of the high-velocity layer is too shallow to be a bedrock surface in rock glaciers, which are mostly thicker than 10 m. In contrast, the latter cases can be attributed to the absence of permafrost or the possible presence of relict permafrost below a thick (>5 m) supra-permafrost talik. In general, the thicknesses of the low-velocity layer distinguish non-vegetated rock glaciers from vegetated ones.

On the other three rock glaciers (PS4, NN8, C2) the thickness of the low-velocity layer ranges from 5 to 9 m. The thicknesses of the low-velocity layer of PS4 bouldery and NN8 pebbly rock glaciers are much thinner than the apparent debris thickness indicated by the heights of the frontal part (20–25 m high), whereas that of C2 vegetated rock glacier approximates the apparent thickness (*c.* 10 m). Bedrock is exposed below the frontal slope of C2. The results for PS4

Table 9.1. *P-wave and DC resistivity stratigraphies of rock glaciers near the lower limit of mountain permafrost*

| Site | P-wave stratigraphy[a] | | | DC resistivity stratigraphy[a] | | | | |
| | Upper layer | | Lower layer | First layer | | Second layer | | Third layer |
	V (m s^{-1})	D (m)	V (m s^{-1})	ρ (kΩ m)	D (m)	ρ (kΩ m)	D (m)	ρ (kΩ m)
Bouldery rock glaciers								
M1	350	1.3 ± 0.3[d]	3700	23	2.9	170	22	1.1
NS2	560	2.8 ± 0.8[d]	2700	52	8.9	84	24	8.0
O2				18	5.4	110	11	0.77
O4	650	2.7 ± 0.6[d]	2100					
PN1	550	2.6 ± 0.5[d]	4300					
PN5	650	1.5 ± 0.7[d]	3800	18	2.2	410	18	0.18
PS3[b]	330	3.1 ± 0.3[d]	4400	4.6–30–9.2	6.5	190	18	7.2
PS4[b]	380	8.8–9.2	2200	0.96	0.42	45	13	0.17
Pebbly rock glaciers								
BNU	390	2.5 ± 0.6[d]	2800	0.60	0.48	2.1–3.7	16	0.90
BNL	370	2.0 ± 0.3[d]	3000	5.7–1.8	3.1	3.7	42	0.26
BN2[c]				3.8	0.2	24	1.4	2.2
BN3U	350	2.4 ± 0.3[d]	2900	1.9–1.0	6.3	2.3	17	1.3
BW1	330	1.6 ± 0.4[d]	2600	0.84	2.2	0.74	9.8	0.16
BW2	350	2.3–5.3	2000	2.5	1.3	0.46	17	1.9
NN2	420	4.6 ± 0.5[d]	2200	2.1	2.9	1.4	6.3	1.9
NN8	450	4.8–8.6	3200	2.0	1.1	1.3	11	2.0
NN10				2.0	0.66	1.3	3.5	5.4
NN11				3.7	2.7	6.4	26	0.62
NN12	340	2.1 ± 0.3[d]	3100	7.4–3.0	3.8	7.7	29	0.022

Densely vegetated (probably relict bouldery) rock glaciers

C1	360–670	>18[e]		1.5	0.86	4.1	8.4	10
C2	410	4.7–6.3	2300	1.1	0.33	4.3	4.3	2.6
NS4				2.9	0.27	81–23	8.4	1.4
NS5	400–730	17–22	3100	1.2	0.39	49	2.8	0.82
PN3	480–700	6.7–12	1600	1.7	0.54	11	2.4	2.2–3.7
PS1	410–950	>15[e]		0.16	0.48	1.4–2.0	24	0.88
S1	410–700	>16[e]						

[a] V: P-wave velocity, D: depth of the layer base, ρ: calculated resistivity.
[b] Electrodes were put on fine materials to decrease the contact resistivity, although boulders dominate on the rock glacier surface.
[c] A thin openwork bouldery layer covers matrix-supported pebbles and cobbles.
[d] Average depth and deviation calculated from the reciprocal method.
[e] Minimum depth assuming that the velocity of the lower layer is $2000 \, \text{m s}^{-1}$.

141

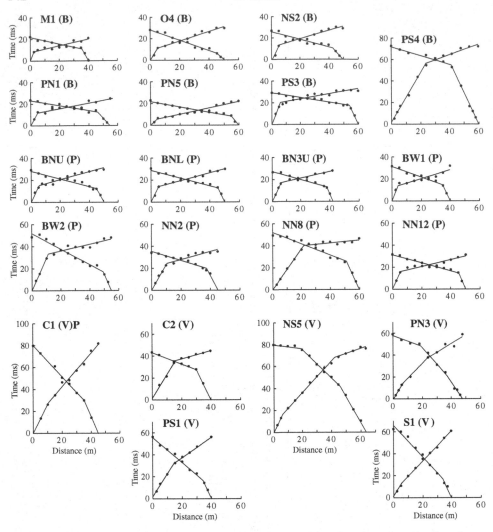

Figure 9.1. Traveltime curves of seismic soundings for three types of rock glaciers. B: bouldery type, P: pebbly type, V: vegetated type.

and NN8 are likely to indicate the presence of permafrost below a supra-permafrost talik whereas the results for C2 indicate the absence of permafrost within the thin rock glacier having a subsided form.

These results and interpretations agree with those of previous studies in which a high-velocity layer (2400–4000 m s^{-1}) in a rock glacier is regarded as frozen where it underlies a thin low-velocity layer (250–1500 m s^{-1}) (e.g. Haeberli and Patzelt 1982, Barsch 1996). P-wave velocities of 2000–2300 m s^{-1}, which are lower than the above-mentioned values for permafrost in rock glaciers, also fall within the values for various types of permafrost (1500–4700 m s^{-1}) (Hunter

1973; see also Chapter 3 and Figure 3.1). In addition, the calculated depths of the permafrost table in M1 and BNU coincide with the directly observed depths within an error of ± 1 m.

9.4.2 DC resistivity

The VES surveys yielded resistivity stratigraphies characteristic of the three types of rock glaciers: bouldery rock glaciers typically show resistivities of 10–50 kΩ m near the surface (<2 m deep in this study) and 10–500 kΩ m in the subsurface (5–10 m deep), whereas pebbly and vegetated rock glaciers typically show resistivities of 1–5 kΩ m near the surface and 0.5–10 kΩ m in the subsurface (Table 9.1). The observed subsurface resistivities within the bouldery rock glaciers are within a range of previously reported values for permafrost in (bouldery) rock glaciers (e.g. Haeberli and Vonder Mühll 1996). Pebbly rock glaciers showed significantly lower resistivities than the values reported previously. The apparent resistivity typically increases from the surface to the subsurface in bouldery and vegetated rock glaciers, whereas it remains nearly constant with a slight increase or decrease in pebbly rock glaciers (Figure 9.2).

The resistivity of the surface layer originates from:

(i)   unfrozen boulders for bouldery rock glaciers,
(ii) unfrozen pebbles/cobbles filled with sandy/silty matrix for pebbly rock glaciers and
(iii) vegetated soil for vegetated rock glaciers.

Exceptionally high or low resistivities of the surface layer can be explained as follows. Two bouldery rock glaciers (PS3, PS4) exhibit anomalously low resistivities (1–5 kΩ m) within the uppermost 0.7 m, because the electrodes were put into fine debris to decrease the contact resistivity. Comparatively high resistivities (24 kΩ m) on BN2 pebbly rock glacier result from a thin matrix-free bouldery layer covering a layer composed of pebbles and cobbles. Two vegetated rock glaciers (NS4, NS5) have thin high resistivity layers (50–80 kΩ m) near the surface (0.3–3 m deep), which indicates the presence of buried matrix-free boulders as confirmed by excavation at NS5.

ERT inversion results also showed that pebbly rock glaciers can be distinguished from bouldery rock glaciers by DC resistivity values of about 10 kΩ m (Figures 9.3 and 9.4). In addition, 2D ERT surveys are highly advantageous for displaying structures under rugged mountain slopes by modelling the lateral heterogeneity of the ground. Resistivities in bouldery rock glaciers show a large spatial variation (100–5000 kΩ m in PN5, 10–300 kΩ m in PS3) (Figure 9.3). The maximum resistivity value obtained from ERT on PN5 is ten times higher than that from the corresponding VES, whereas the VES on PS3 appears to detect the maximum

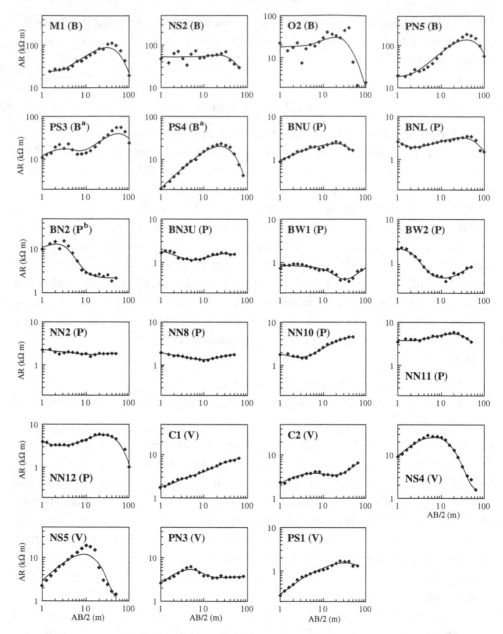

Figure 9.2. Apparent resistivity curves for three types of rock glaciers. B: bouldery type, P: pebbly type, V: vegetated type. The abscissa is given by the half-length of the measurement profile (AB/2) and the ordinate by apparent resistivity (AR). Note the different scale of apparent resistivity for each graph. [a]Electrodes were put on fine materials to decrease the contact resistivity, although boulders dominate on the rock glacier surface. [b]A thin openwork bouldery layer covers matrix-supported pebbles and cobbles.

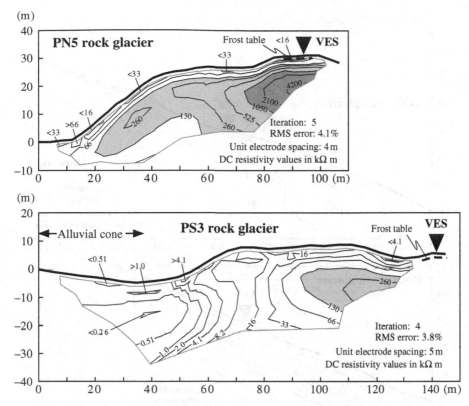

Figure 9.3. ERT inversion results for bouldery rock glaciers. VES indicates the position of a one-dimensional sounding profile placed perpendicularly to the flow line. The depth of the frost table was estimated by seismic soundings.

resistivity value indicated by the 2D ERT survey (Table 9.1, Figure 9.3). The large difference in the calculated resistivities between VES and ERT surveys on PN5 is mainly attributed to the limitations of 1D inverse modelling of an increasing apparent-resistivity curve close to the theoretical maximum angle (45°), which originates from an increase in DC resistivity of several orders of magnitude within a relatively short distance (further possible causes are described in Chapter 1).

The observed resistivities (0.25–20 kΩ m) in the longitudinal profiles of the pebbly rock glaciers show a similar range to those obtained from the VES surveys (Table 9.1, Figure 9.4). The calculated resistivities from ERT surveys on BNU, BNL, BW1 and BW2 roughly correspond to those from the VES surveys below the intersections of the measurement profiles. In addition, elliptic layers with relatively high resistivities are distinguished in the ERT profiles and the high resistivities in BN3U (4–9 kΩ m), BW1 (2–6 kΩ m) and NN12 (10–20 kΩ m) were not detected by the 1D surveys (Table 9.1, Figure 9.4).

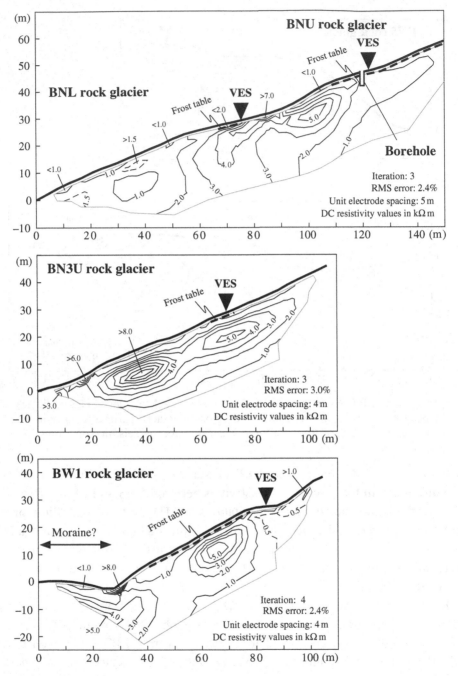

Figure 9.4.   ERT inversion results for pebbly rock glaciers. VES indicates the position of a one-dimensional sounding profile placed perpendicularly to the flow line. The depth of the frost table was estimated by seismic soundings.

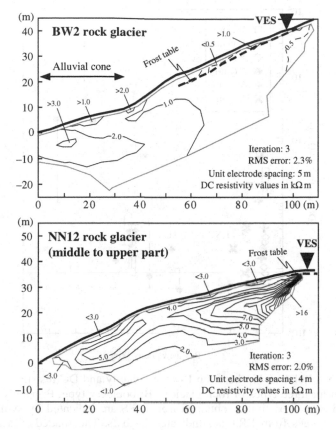

Figure 9.4.   (cont.)

The directly observed structures in M1 and BNU show that the subsurface resistivities largely depend on the grain size and ice supersaturation of the permafrost. The highly ice-supersaturated permafrost in M1 results in an apparent resistivity value of $100\,k\Omega\,m$ (Figure 9.2), whereas specific resistivity values between $300{-}1000\,k\Omega\,m$ were given by ERT (Hauck and Vonder Mühll 2003; Chapter 1). In contrast, the slightly ice-supersaturated debris at the melting point in BNU shows a resistivity of $2\,k\Omega\,m$, which is much lower than the typical values for the permafrost in bouldery rock glaciers (Table 9.1, Figure 9.4).

9.5 Discussion

Permafrost in the studied rock glaciers appears to have a mean annual temperature higher than $-2\,°C$, because M1 is located at the coldest site with a permafrost temperature mostly above $-2\,°C$ (Ikeda 2006). Thus, all geophysical properties discussed below originate from such warm permafrost.

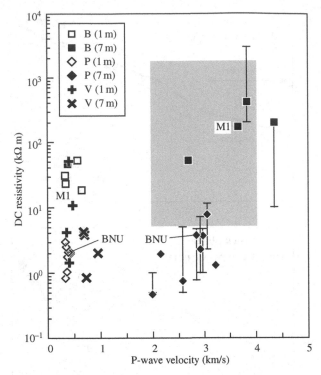

Figure 9.5.   Relationship between P-wave velocity and DC resistivity at 1 and 7 m depth for three types of rock glaciers. B: bouldery type, P: pebbly type, V: vegetated type. Resistivities obtained from VES are indicated by symbols and resistivity ranges from ERT are indicated by bars. The shaded area indicates typical values for permafrost in (bouldery) rock glaciers (Haeberli and Vonder Mühll, 1996).

Figure 9.5 displays the relationship between P-wave velocity and DC resistivity at near-surface (1 m) and core (7 m) depths for the three types of rock glaciers. The values for a thin vegetated rock glacier (C2) were eliminated from the figure because bedrock may exist below 5 m. The core P-wave velocity of the non-vegetated rock glaciers ($\geq 2000\,\mathrm{m\,s^{-1}}$), except for PS4 which may have supra-permafrost talik, is much higher than the near-surface velocity (330–650 m s^{-1}). In contrast, the difference in velocity between near-surface (360–480 m s^{-1}) and core (670–950 m s^{-1}) is small for the vegetated rock glaciers. Thus, conventional seismic soundings are probably sufficient to indicate permafrost presence in rock glaciers thicker than 10 m, because the vertical difference in continuity and density between frozen and unfrozen debris is much larger than the lateral heterogeneity within a short distance from the sounding (*c*. 50 m). Hence, analysis of the presence or absence of permafrost will be based on the seismic soundings in the following.

In contrast to P-wave velocity, DC resistivity has a wide, overlapping range ($1-100\,k\Omega\,m$) between the unfrozen near-surface and the frozen core, if the permafrost is warmer than $-2\,°C$ (Figure 9.5). In particular, pebbly rock glaciers rarely exhibit a difference concerning the DC resistivity values of the surface and subsurface layers. For example, the permafrost, indicated by the seismic soundings in BNU, BW1 and BW2, may be difficult to detect based on resistivity values alone, even when using ERT (Figure 9.4). The decreasing apparent resistivity curves of BW1, BW2, NN2 and NN8 do not indicate the presence of permafrost either (Figure 9.2). Similarly, no distinction concerning subsurface resistivity values is possible between pebbly (active/inactive) and vegetated (relict) rock glaciers (Figure 9.5). The increasing resistivity curve of C1 vegetated (relict) rock glacier is similar to that of BNU active pebbly rock glacier (Figure 9.2). In addition, the resistivity of matrix-free (unfrozen) boulders, reaching up to $50\,k\Omega\,m$ in this study, occasionally makes it difficult to distinguish between an active layer and a permafrost layer in bouldery rock glaciers (e.g. NS2). Hence, where a high-resistivity layer ($>100\,k\Omega\,m$) is absent, DC resistivity methods are less consistent to identify permafrost in rock glaciers than seismic methods.

Different temperatures between unfrozen and frozen layers in a rock glacier do not always result in DC resistivity differences. This contrasts with the observation that DC resistivity of saturated sand and gravel sharply increases when ground temperatures fall below $0\,°C$ (Hoekstra and McNeill 1973; see also Chapter 7). However, the analogy from their result is misleading for interpreting DC resistivity in rock glaciers because the active layer is dry rather than water-saturated (i.e. there is a difference in water content between the active layer and permafrost).

The unfrozen water content in frozen debris depends on the total surface area of soil particles as well as on the temperature. Thus, the large difference in DC resistivity of permafrost between bouldery and pebbly rock glaciers partly reflects the material composition, because fine-rich pebbly rock glaciers hold more unfrozen water than fine-poor bouldery rock glaciers even below the melting point (see Ikeda and Matsuoka 2006 for the material composition). Similarly, the difference in DC resistivity between active/inactive and relict bouldery rock glaciers is attributed to material composition, because degradation of permafrost eventually results in weathering of clasts and subsequent production of fine debris, probably reducing the resistivity in rock glaciers (Ikeda and Matsuoka 2002).

DC resistivity is also a useful indicator of ice content (especially supersaturation) in warm permafrost (0 to $-2\,°C$). A wide variation in ice contents from less than 60% in ice-cemented debris (e.g. Fisch *et al.* 1977, Elconin and LaChapelle 1997, Arenson *et al.* 2002) to nearly 100% in massive ice (e.g. Vonder Mühll and Holub 1992, Arenson *et al.* 2002) has been reported for bouldery rock glaciers. Ice supersaturation increases the resistivity of permafrost because of a decrease in the continuity of unfrozen water films lapping the soil particles (Fortier *et al.* 1994). Thus, resistivities higher than those of matrix-free boulders (>100 kΩ m) probably indicate the presence of a highly ice-supersaturated layer (Haeberli *et al.* 1998). The very high values (1000–5000 kΩ m) in PN5 indicate the presence of massive ice (>10 m thick) even at relatively low temperatures as in the upper part of the M1 rock glacier (Hauck and Vonder Mühll 2003).

9.6 Conclusions

Near the lower limit of mountain permafrost, the DC resistivity of rock glaciers reflects structural differences (e.g. bouldery or pebbly, ice-cemented or highly ice-rich) rather than thermal differences (frozen or unfrozen). Where permafrost is detected by other properties (e.g. P-wave velocity), DC resistivity indicates the grain-size distribution and ice content (see also Chapter 1).

The DC resistivity of permafrost differs significantly between bouldery (10–5000 kΩ m) and pebbly (0.5–20 kΩ m) rock glaciers. Resistivities higher than 100 kΩ m indicate a highly ice-supersaturated layer in bouldery rock glaciers, whereas resistivities of pebbly rock glaciers indicate ice-cemented pebbles/cobbles with relatively high unfrozen water content.

The wide, overlapping range of DC resistivity values between non-vegetated and vegetated rock glaciers and between the near-surface and subsurface layer of rock glaciers demonstrates that the absence of permafrost is difficult to predict from DC resistivity alone on heterogeneous mountain slopes. Thus, previous studies mapping mountain permafrost based on DC resistivity methods alone may require further improvement.

Acknowledgments

This study was financially supported by Grant-in-Aid for JSPS Fellows from the Ministry of Education, Science and Culture, Japan and by a Sasakawa Scientific Research Grant from the Japan Scientific Society. The author acknowledges N. Matsuoka for academic and logistical support, F. Keller for logistical help, K. Hirakawa, T. Watanabe, M. Aoyama, K. Fukui, S. Iwasaki for cooperation in the field, Mi. Abe, Ma. Abe, T. Date and Y. Eda for field assistance.

REFERENCES

Arenson, L., Hoelzle, M. and Springman, S. (2002). Borehole deformation
    measurements and internal structures of some rock glaciers in Switzerland.
    Permafrost Periglacial Processes, **13**, 117–135.

Barsch, D. (1996). *Rockglaciers: Indicators for the Present and Former Geoecology in High Mountain Environments*. Springer.

Elconin, R. F. and LaChapelle, E. R. (1997). Flow and internal structure of a rock glacier. *Journal of Glaciology*, **43**, 238–244.

Fisch, W. Sr., Fisch, W. Jr. and Haeberli, W. (1977). Electrical D. C. resistivity soundings with long profiles on rock glaciers and moraines in the Alps of Switzerland. *Zeitschrift für Gletscherkunde und Glazialgeologie*, **13**, 239–260.

Fortier, R., Allard, M. and Seguin, M. K. (1994). Effect of physical properties of frozen ground on electrical resistivity logging. *Cold Regions Science and Technology*, **22**, 361–384.

Haeberli, W. (1985). *Creep of Mountain Permafrost: Internal Structure and Flow of Alpine Rock Glaciers*. Mitteilungen der Versuchsanstalt für Wasserbau, Hydrologie und Glaziologie, 77, 142pp.

Haeberli, W. and Patzelt, G. (1982). Permafrostkartierung im Gebiet der Hochebenkar-Blockgletscher, Obergurgl, Ötztaler Alpen. *Zeitschrift für Gletscherkunde und Glazialgeologie*, **18**, 127–150.

Haeberli, W. and Vonder Mühll, D. (1996). On the characteristics and possible origins of ice in rock glacier permafrost. *Zeitschrift für Geomorphologie, Supplement*, **104**, 43–57.

Haeberli, W., Huder, J., Keusen, H.-R., Pika, J. and Röthlisberger, H. (1988). Core drilling through rock glacier permafrost. *Proceedings of the 5th International Conference on Permafrost*, Trondheim, Norway, 937–942.

Haeberli, W., Hoelzle, M., Keller, F., Vonder Mühll, D. and Wagner, S. (1998). Ten years after the drilling through the permafrost of the active rock glacier Murtèl, eastern Swiss Alps: answered questions and new perspectives. *Proceedings of the 7th International Conference on Permafrost*, Yellowknife, Canada, 403–410.

Hauck, C. and Vonder Mühll, D. (2003). Inversion and interpretation of two-dimensional geoelectrical measurements for detecting permafrost in mountainous regions. *Permafrost and Periglacial Processes*, **14**, 305–318.

Hoekstra, P. and McNeill, D. (1973). Electromagnetic probing of permafrost. *Proceedings of the 2nd International Conference on Permafrost*, Yakutsk, Siberia, 517–526.

Hunter, J. A. M. (1973). The application of shallow seismic methods to mapping of frozen surficial materials. *Proceedings of the 2nd International Conference on Permafrost*, Yakutsk, Russia, 527–535.

Ikeda, A. (2006). Combination of conventional geophysical methods for sounding the composition of rock glaciers in the Swiss Alps. *Permafrost and Periglacial Processes*, **17**, 35–48.

Ikeda, A. and Matsuoka, N. (2002). Degradation of talus-derived rock glaciers in the Upper Engadin, Swiss Alps. *Permafrost and Periglacial Processes*, **13**, 145–161.

Ikeda, A. and Matsuoka N. (2006). Pebbly versus bouldery rock glaciers: morphology, structure and processes. *Geomorphology*, **73**, 279–296.

Ikeda, A., Matsuoka, N. and Kääb, A. (2008). Fast deformation of pevennially frozen debris in a warm rock glacier in the Swiss Alps: An effect of liquid water. *Journal of Geophysical Research*, **113**, F01021.

Ishikawa, M., Watanabe, T. and Nakamura, N. (2001). Genetic differences of rock glaciers and the discontinuous mountain permafrost zone in Kanchanjunga Himal, eastern Nepal. *Permafrost and Periglacial Processes*, **12**, 243–253.

King, L., Fisch, W., Haebrli, W. and Wächter, H. P. (1987). Comparison of resistivity and radio-echo soundings on rock glacier permafrost. *Zeitschrift für Gletscherkunde und Glazialgeologie*, **23**, 77–97.

Palmer, D. (1986). *Refraction Seismics*. Geophysical Press, London.
Vonder Mühll, D. S. and Holub, P. (1992). Borehole logging in alpine permafrost, Upper Engadin, Swiss Alps. *Permafrost and Periglacial Processes*, **3**, 125–132.
Vonder Mühll, D. S., Hauck, C. and Lehmann, F. (2000). Verification of geophysical models in Alpine permafrost by borehole information. *Annals of Glaciology*, **31**, 300–306.
Vonder Mühll, D., Hauck, C. and Gubler, H. (2002). Mapping of mountain permafrost using geophysical methods. *Progress in Physical Geography*, **26**, 643–660.

10

Quantifying the ice content in low-altitude scree slopes using geophysical methods

C. Hauck and C. Kneisel

10.1 Introduction

In many periglacial applications ground ice detection and quantification is the major objective. Typical examples are low-altitude scree slopes, which are covered with blocky material and may contain ground ice throughout the year far below the regional limit of permafrost due to microclimatic conditions that resemble those of high-altitude periglacial areas (see Sawada *et al.* 2003, Delaloye *et al.* 2003, Delaloye and Lambiel 2005, Zacharda *et al.* 2007). Essential preconditions for this extraordinary microclimatic phenomenon are assumed to be a thick layer of blocks with an open void system, i.e. steep slopes with almost no fine material. Summer ice observations and cold air outflow from the blocks in the near subsurface, as well as the occurrence of cold-adapted mosses and different invertebrate groups (e.g. beetles and spiders) that normally live in high alpine or polar areas are used as indicators for the possible presence of ground ice (Gude *et al.* 2003, Zacharda *et al.* 2005). The 'thermal semi-conductor' effect of the coarse rocks is also widely used in construction, e.g. for the permanent cooling of railroad embankments in permafrost regions (Cheng *et al.* 2007).

However, whether significant ice occurrences could be permanently present within the scree slopes is still an open question. Delaloye and Lambiel (2005) describe a cold ventilated talus slope in the Swiss Alps, where sediments more or less saturated with ice were found at 7–8 m depth during construction of two cable car pylons. Sawada *et al.* (2003) detected ground ice at a depth of 1.5–1.6 m throughout the year within a scree slope on Hokkaido Island, Japan, but this cannot be generalised to other scree slopes due to the large differences in aspect, altitude and weathering characteristics between the various scree slopes. Geophysical methods are sometimes used to investigate the possible ground ice

Applied Geophysics in Periglacial Environments, eds. C. Hauck and C. Kneisel. Published by Cambridge University Press. © Cambridge University Press 2008.

occurrences, but the results of single methods are often inconclusive due to the large air voids between the blocks and the comparatively small ice contents (see Gude *et al.* 2003). In this contribution a newly developed model for quantification of ice, water and air content using data from ERT and refraction seismic tomography (Hauck *et al.* 2005, Hauck *et al.* 2008) is applied to several low-altitude scree slopes in Central Europe.

10.2 Methods

In partly or permanently frozen ground, subsurface material may consist of four different phases: two solid phases (rock/soil matrix as well as ice), a liquid (unfrozen pore water) and a gaseous phase (air-filled pore space and cavities). Except for the analysis of borehole data, the composition of the subsurface material can only be inferred through indirect geophysical investigations. Due to their complementary characteristics, geoelectric and seismic methods are often combined to avoid ambiguities in the interpretation of the results (see Kneisel and Hauck 2003, and Chapter 9.). In particular, signal propagation occurs usually within the liquid phase for electrical methods (through ionic conduction within the pore water), whereas it occurs in the solid phase (rock or ice) for seismic methods. A combination of both methods, ideally as tomographies along a common two- or three-dimensional grid, is therefore expected to enable distinction of the different phases in the subsurface.

The indirect nature of geophysical soundings requires a relation between the measured variable (electrical resistivity, seismic velocity) and the respective parts of the material composition (rock/soil, water, air, ice). To determine the actual volumetric fractions of the four phases (namely the ice content, air content, water content and porosity), Hauck *et al.* (2008) proposed a so-called four-phase model based on two-dimensional ERT and refraction seismic velocity data sets. The model was originally tested on two Alpine rock glaciers with spatially variable ice content, and validation was obtained through a series of boreholes.

The four-phase model is based on two well-known geophysical mixing rules for electrical resistivity and seismic P-wave velocity (Archie 1942, Timur 1968). The observed resistivity and velocity data are used as model input on a two-dimensional grid; material properties such as resistivity and P-wave velocity of the host rock material and the pore water have to be known beforehand. In the basic formulation used for this case study, the model consists of a system of three equations and four unknown volumetric contents. Consequently, one of the volumetric contents has to be specified. For this case study a constant porosity ($= 1 -$ rock content) for each scree slope was assumed. In addition, model computations for a 'no-ice case' (ice content $= 0$) were conducted, yielding a variable porosity model in addition to the models for air and water content. By this, it could be evaluated whether (i) the constant porosity approach and/or (ii) the no-ice approach lead to realistic estimations of the volumetric contents. In a further step the remaining free model parameters can be determined by a quasi Monte-Carlo approach, the results of which are used additionally as an indicator of the reliability of the model results. Details of the model set-up and its sensitivity to the various free parameters can be found in Hauck *et al.* (2008).

Acquisition of seismic and resistivity data was conducted using standard multi-channel geoelectric and seismic instruments (see Chapters 1 and 3). Typically 12 geophones with 5 m spacing and 36 electrodes with 5 m spacing were used, resulting in an almost doubled investigation volume for ERT compared to the seismic survey. Data acquisition was hindered by the presence of very dry and loosely connected blocks at the scree surface, where sensor coupling is even more difficult than on rock glaciers. Inversion of resistivity and seismic data was performed using RES2DINV and a refraction tomographic inversion scheme introduced by Lanz *et al.* (1998), respectively. Inversion parameters were chosen according to the guidelines presented in the Chapters 1 and 3, respectively.

10.3 Field sites

Results from four different scree slopes are presented within this case study: (i) Val Bever (46° 33′ 20″ N/9° 51′ 30″ E, eastern Swiss Alps, 1800 m a.s.l.); (ii) Präg (47° 46′ 55″ N/7° 57′ 47″ E, Black Forest, southwestern Germany, 720 m a.s.l.); (iii) Zastler (47° 55′ 05″ N/7° 59′ 30″ E, Black Forest, southwestern Germany, 590 m a.s.l.) and (iv) La Glacière (48° 06′ 25″ N/6° 58′ 00″ E, Vosges Mountains, eastern France, 680 m a.s.l.).

The Val Bever field site may be characterised as an alpine scree slope, but is situated below the timberline and the regional limit of discontinuous permafrost in the Swiss Alps. Most parts of the slope are covered with vegetation (grass and moss). Permafrost was detected indirectly in a number of consecutive studies (Kneisel *et al.* 2000, Kneisel and Hauck 2003, Hauck and Kneisel 2006). Direct confirmation of perennial frozen ground was obtained through a borehole drilled and instrumented in 2006, where subzero temperatures are recorded below a depth of 2–3 m (Kneisel 2007).

The three non-alpine scree slopes at Präg, Zastler and La Glacière have only sparse vegetation cover and exhibit a distinct open void system between the blocks. Evidence for the specific microclimatic conditions enabling seasonal or even perennial ice presence was first found by Molenda (1996); however, from these results no indications of the presence of significant ground ice occurrences can be deduced. In recent years, observation of summer ice occurrences near the surface has ceased at all three non-alpine scree slopes. Whether this may be an effect of changes in the local climate conditions, as was reported by Tanaka *et al.* (1999) for a scree slope in Milyang, Korea, has not been investigated so far.

ERT and seismic surveys were conducted in summer 2002 at the three non-alpine scree slopes. At all three sites only one cross-profile at the foot of the scree slope was conducted. It should be kept in mind that this single profile cannot adequately represent the whole scree slope; however, its location was chosen based on the largest probability for the occurrence of ground ice. At Val Bever, longitudinal surveys have been conducted repeatedly since 1998; here, we will show the four-phase model results based on the initial geophysical measurements from summer 1998 (Kneisel and Hauck 2003).

10.4 Results

Figure 10.1 (Plate 10) shows the results of the electrical resistivity and seismic velocity inversion models as well as the results from the four-phase model for La Glacière. Velocities are comparatively small ($<2000\,\text{m/s}$) and resistivities are high (between 10 and $100\,\text{k}\Omega\,\text{m}$) throughout most of the model domain. Even though some areas with anomalously high or low resistivity and velocity values can be identified, the overall distribution is non-trivial and the delineation of regions with enhanced ice, water and air content is not easily deduced from the tomograms alone.

The four lower panels in Figure 10.1 show the calculated ice, water and air contents using the four-phase model assuming a homogeneously distributed porosity of 40% (simulating the large air-, water- or ice-filled voids between the boulders) over the whole model domain. For the calculation the predefined material constants shown in Table 10.1 were used. The calculated porosity model for the no-ice assumption shows porosity values between 30 and 50% (except for the uppermost metre), making 40% a reasonable assumption as a mean porosity value for the constant porosity model. Naturally, no kind of porosity model will be able to correctly represent the highly heterogeneous subsurface conditions assumed to be present within a scree slope. However, results from this porosity-dependent model may be interpreted as indicating to what extent the available pore space is filled with the respective phases, independent of the absolute value and the distribution of the porosity within the scree slope. In a further step, a porosity-independent version of the four-phase model, where the two-dimensional distribution of porosity is estimated using a Monte Carlo approach, can be applied.

The calculated ice content within La Glacière scree slope is low except for a small region between horizontal distances 20 and 40 (below a small depression)

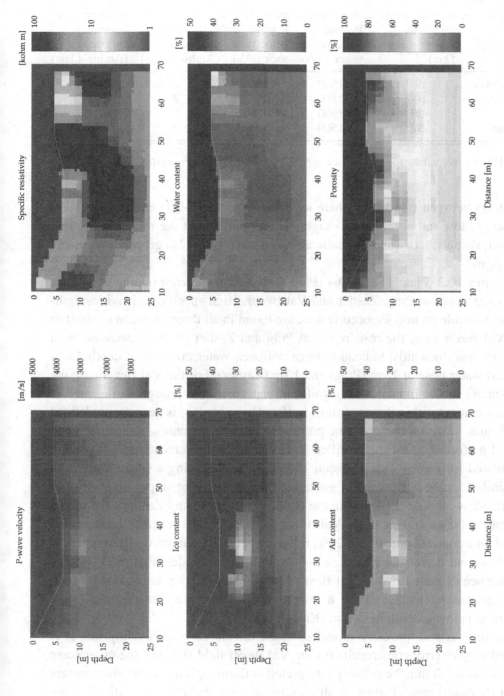

Figure 10.1. P-wave velocity and specific resistivity inversion results for La Glacière scree slope in the Vosges Mountains, eastern France. The lower four panels show calculated four-phase model results for the ice, water and air content as well as porosity. Note that ice, water and air content were calculated with an assumed constant porosity of 40%, whereas the variable porosity model was calculated for an assumed ice content of 0%. The white line marks the surface topography. For colour version see Plate 10.

Table 10.1. *Material constants used for the four-phase model calculations in this study*

	Resistivity of pore water (Ω m)	P-wave velocity water (m/s)	P-wave velocity rock (m/s)	Cementation exponent m (Archie's law)	Saturation exponent n (Archie's law)
Val Bever	50	1500	4500	2	2
Le Glacière	100	1500	4000	2	2
Präg	30	1500	5200	2	2
Zastler	32	1500	5250	2	2

The P-wave velocities for air and ice are 330 m/s and 3500 m/s, respectively.

at depths between 1 and 5 m, where ice contents up to 15–20% are found. Water content is low and air content is high ($\sim$35%) throughout the model indicating, not surprisingly, a dry and basically air-filled scree slope. No ground water table was found.

Figures 10.2 (Plate 11) to 10.4 (Plate 13) show the corresponding results for the scree slopes at Präg, Zastler and Val Bever. High specific resistivities, indicating possible ground ice occurrences, are found in all three cases. In contrast to the Val Bever case, the resistivities at Präg and Zastler strongly decrease from 8–10 m depth onwards, indicating large unfrozen water contents at depth, i.e. a ground water table at 15 m (Präg) and 10 m (Zastler). For the Val Bever case the four-phase model indicates no significant water content throughout the profile. The air content within the uppermost 5–10 m (large blocks) is near the predefined maximum value of the constant porosity model in all cases, indicating good model performance for the air-filled blocky layer of the scree slopes. Maximum calculated ice contents are between 15 and 25%, indicating a surprisingly high possibility for the occurrence of ground ice within the scree slopes. In contrast to the two low-altitude field sites in the Black Forest (Präg and Zastler), the model indication for ground ice at the Val Bever field site is even stronger, because it is based on a constant porosity of 20%, resulting in full ice saturation (ice content = 20%) around horizontal distances 0–50 m within the model (Figure 10.4). The occurrence of permafrost at Val Bever was recently directly confirmed by temperature measurements within a borehole where subzero temperatures were recorded below a depth of 2–3 m (Kneisel 2007).

As the variable porosity model is based on the zero-ice assumption, it is bound to give erroneous results for the Val Bever field site. For the other scree slopes, its result may be reliably interpreted in those regions, where the constant porosity model predicts only small ice contents. For the three low-altitude scree slopes La Glacière, Präg and Zastler the porosity results indicate a porosity

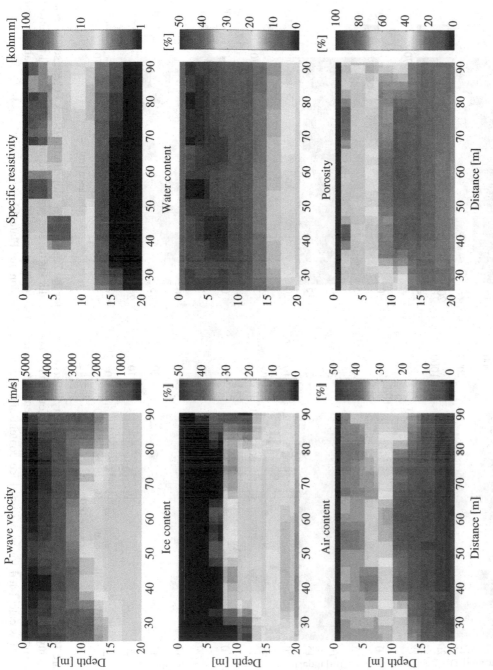

Figure 10.2. P-wave velocity and specific resistivity inversion results for the Präg scree slope in the Black Forest, southwestern Germany. The lower four panels show calculated four-phase model results for the ice, water and air content as well as porosity. Note that ice, water and air content were calculated with an assumed constant porosity of 40%, whereas the variable porosity model was calculated for an assumed ice content of 0%. For colour version see Plate 11.

159

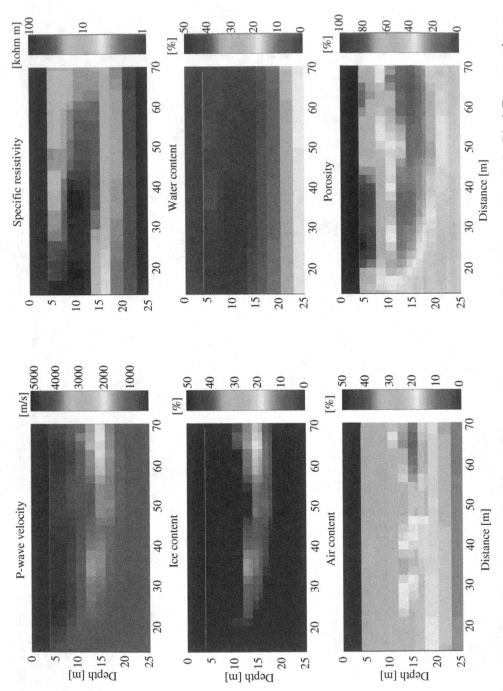

Figure 10.3.   P-wave velocity and specific resistivity inversion results for the Zastler scree slope in the Black Forest, southwestern Germany. The lower four panels show calculated four-phase model results for the ice, water and air content as well as porosity. Note that ice, water and air content were calculated with an assumed constant porosity of 30%, whereas the variable porosity model was calculated for an assumed ice content of 0%. For colour version see Plate 12.

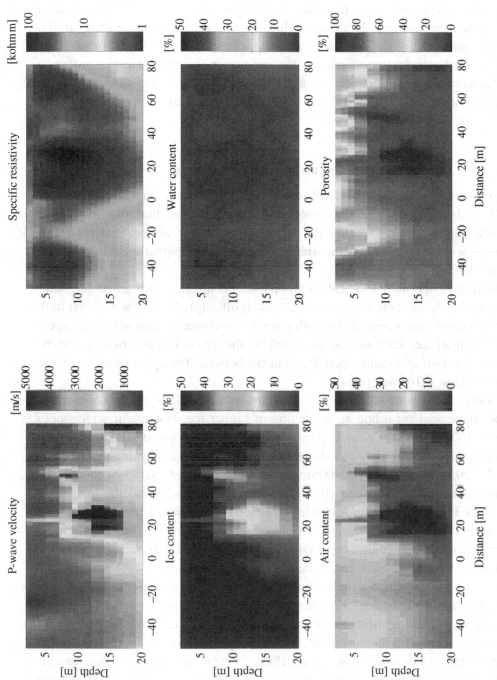

Figure 10.4. P-wave velocity and specific resistivity inversion results for the Val Bever scree slope in the Upper Engadine, eastern Swiss Alps. The lower four panels show calculated four-phase model results for the ice, water and air content as well as porosity. Note that ice, water and air content were calculated with an assumed constant porosity of 20%, whereas the variable porosity model was calculated for an assumed ice content of 0%. For colour version see Plate 13.

significantly larger than 50% in the upper block layer and values between 30 and 40% below. However, from the results it is seen that ice and rock content are difficult to distinguish, as both velocity and resistivity can be quite similar for ice and rock for the specific conditions present in scree slopes. In addition, the limited penetration depth of the refraction seismic surveys (using a sledgehammer as source, see Chapter 3) inhibited the reliable detection of the bedrock layer and restricted the analysis to the upper 20 m (see also Otto and Sass 2006).

10.5 Discussion and conclusions

First results of a newly developed four-phase model based on a combination of ERT and refraction seismic tomography data indicate that this method is indeed capable of computing the spatial distribution of ice, water and air content within the unique geomorphological and microclimatic environment of scree slopes.

Results of the model indicate a ground ice occurrence with saturated conditions at the alpine scree slope, which is still significantly below the regional limit of permafrost, and a possible indication for the existence of ground ice within the three non-alpine scree slopes. In addition, the ground water table could be detected in two cases and a porosity gradient between the upper block layer and depths below 10 m was found.

Comparisons with other studies on alpine and non-alpine scree slopes show good agreement regarding an estimation of higher ice contents in scree slopes above 1500 m a.s.l. (e.g. at Lapires/Swiss Alps, Delaloye and Lambiel 2005) compared to low-altitude scree slopes (e.g. within the Bohemian highlands, Gude et al. 2003, Zacharda et al. 2005). It may be speculated that even though the temperature gradient between the near-surface atmosphere and the interior of the scree is large for the low-altitude scree slopes, minimum temperatures within the screes are not low enough to generate or preserve large ice occurrences. Furthermore, at higher altitudes ground ice occurrences within screes can also be generated by other means, e.g. buried ice or snow patches.

Due to the described deficiencies of the four-phase model (the necessity to predefine a porosity model, dependence on partly unknown material properties, difficulties in distinguishing between ice and bedrock occurrences), the results should be seen as indications of the relative proportions of the four phases rather than reliable estimations of absolute values. Some of these deficiencies are currently being improved; however, a critical glaciological and geomorphological assessment of the model results is always recommended.

REFERENCES

Archie, G. E. (1942). The electrical resitivity log as an aid in determining some reservoir characteristics, *American Institute of Mining and Metallurgical Engineers*, 55–62.

Cheng, G., Lai, Y., Sun, Z. and Jiang, F. (2007). The 'thermal semi-conductor' effect of crushed rocks. *Permafrost and Periglacial Processes*, **18**, 151–160.

Delaloye, R. and Lambiel, C. (2005). Evidence of winter ascending air circulation throughout talus slopes and rock glaciers situated in the lower belt of alpine discontinuous permafrost (Swiss Alps). *Norwegian Journal of Geography*, **59**(2), 194–203.

Delaloye, R., Reynard, E., Lambiel, C., Marescot, L. and Monnet, R. (2003). Thermal anomaly in a cold scree slope, Creux du Van, Switzerland. *Proceedings of the 8th International Conference on Permafrost*, Zürich, Switzerland, 175–180.

Gude, M., Dietrich, S., Mäusbacher, R., Hauck, C., Molenda, R., Ruzicka, V. and Zacharda, M. (2003). Permafrost conditions in non-alpine scree slopes in central Europe. *Proceedings of the 8th International Conference on Permafrost*, Zürich, Switzerland, 331–336.

Hauck, C. and Kneisel, C. (2006). Application of capacitively-coupled and DC electrical resistivity imaging for mountain permafrost studies. *Permafrost and Periglacial Processes,* **17**(2), 169–177.

Hauck, C., Böttcher, M. and Kottmeier, C. (2005). Quantifying subsurface ice and unfrozen water content using a geophysical monitoring approach. *Geophysical Research Abstracts*, **7**, 07708.

Hauck, C., Bach, M. and Hilbich, C. (2008). A 4-phase model to quantity subsurface ice water content in permafrost vegious based on geophysical data sets. Proceedings of the 9th International Conference on Permafrost, Fairbanks, Alasha, 2008, 6pp.

Kneisel, C. (2007). The nature and dynamics of frozen ground in alpine and subarctic periglacial environments. Unpublished professorial dissertation, Faculty of Geoscience, University of Würzburg.

Kneisel, C. and Hauck, C. (2003). Multi-method geophysical investigation of an isolated permafrost occurrence. *Zeitschrift für Geomorphologie, Supplement*, **132**, 145–159.

Kneisel, C., Hauck, C. and Vonder Mühll, D. (2000). Permafrost below the timberline confirmed and characterized by geoelectrical resistivity measurements, Bever Valley, eastern Swiss Alps. *Permafrost and Periglacial Processes*, **11**, 295–304.

Lanz, E., Maurer, H. R. and Green, A. G. (1998). Refraction tomography over a buried waste disposal site. *Geophysics*, **63**(4), 1414–1433.

Molenda, R. (1996). Zoogeographische Bedeutung Kaltluft erzeugender Blockhalden im außeralpinen Mitteleuropa. Untersuchungen an Arthropoda, insbesondere Coleoptera. *Verhandlungen des naturwissenschaftlichen Vereins Hamburg*, **35**, 5–93.

Otto, J. C. and Sass, O. (2006). Comparing geophysical methods for talus slope investigations in the Turtmann valley (Swiss Alps). *Geomorphology*, **76**, 257–272.

Sawada, Y., Ishikawa, M. and Ono, Y. (2003). Thermal regime of sporadic permafrost in a block slope on Mt. Nishi-Nupukaushinupori, Hokkaido Island, Northern Japan. *Geomorphology*, **52**, 121–130.

Tanaka, H. L., Moon, S.-E. and Hwang, S.-J. (1999). An observational study of summertime ice formation at the ice valley in Milyang, Korea. Science Report, Institute of Geosciences. University of Tsukuba, **20**, 33–51.

Timur, A. (1968). Velocity of compressional waves in porous media at permafrost temperatures. *Geophysics*, **33**(4), 584–595.

Zacharda, M., Gude, M., Kraus, S., Hauck, C., Molenda, R. and Ruzicka, V. (2005). The relict mite Rhagidia gelida (Acari, Rhagidiidae) as biological cryoindicator of periglacial microclimate in European highland screes. *Arctic, Antarctic and Alpine Research*, **37**(3), 402–408.

Zacharda, M., Gude, M. and Ruzicka, V. (2007). Thermal regime of three low elevation scree slopes in central Europe. *Permafrost and Periglacial Processes*, **18**, 301–308.

11

The use of GPR in determining talus thickness and talus structure

O. Sass

11.1 Introduction

Talus deposits are widespread in arctic and alpine environments, and represent an important sediment store in these regions. However, detailed information on total thickness and structure of the sediment bodies is sparse. Ground-penetrating radar offers a promising approach to collect data on loose debris quickly and in high resolution. The damping of radar waves is primarily dependent upon the dielectric constant and on the electrical conductivity of the subsurface. Talus deposits are usually very dry and thus very high-resistive (similar to scree slopes, see Chapter 10). Accordingly, the basic preconditions for GPR measurements are very good.

11.2 Study sites and data acquisition

The exemplary radargrams presented were measured in the Lechtaler Alps and in the Kühtai area (Austria) from 2003 to 2006 (Figure 11.1). The 'Parzinn' is a wide cirque in the Lechtaler Alps at an elevation of 2000–2700 m. Large talus cones have developed under dolostone rockwalls. Most of the cones end down-slope at the reverse angle of late-glacial moraine ridges. In numerous radar profiles, a penetration depth of up to 50 m was achieved. The results were validated by additional geoelectric and seismic measurements (Sass 2006). The Kühtai area lies in the Central Alps at an elevation of 2300–3000 m, the rockwalls consist of gneiss and mica-schist. Two steep talus cones were investigated by GPR and geoelectric longitudinal and cross profiles. One of the cones ended downslope at a small lake.

For the investigations presented, a RAMAC GPR (MALÅ Geosystems, Sweden) was used. The equipment proved suitable for the rough terrain.

Applied Geophysics in Periglacial Environments, eds. C. Hauck and C. Kneisel. Published by Cambridge University Press. © Cambridge University Press 2008.

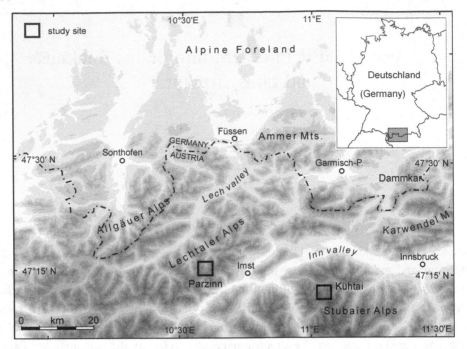

Figure 11.1.   Location of the study sites presented.

However, the mechanical stress on the antennae is very high and parts of the construction could be improved for better handling and stability. It is definitely advisable to bring replacement parts, especially a spare set of glass-fibre cables. Using the novel MALÅ Rough Terrain Antennae (RTA) is a robust alternative option; however, they are not suitable for WARR measurements (see below). A ruggedised field computer should be used because it is very likely that the measuring staff will slip more than once on the loose debris and the laptop might be dropped.

11.3 Results

The propagation velocity was determined in several areas of investigation by performing WARR profiles (measurements with stepwise increasing antenna distance, see Chapter 4). The results range from 0.09 to 0.105 m/ns on moraine ridges and 0.10 to 0.14 m/ns on most talus deposits (Table 11.1). The coarser and drier the talus deposit, the larger is the velocity. Conversely, when a higher proportion of fines (and thus, water) is present, the velocity can drop to 0.09 m/ns or less. It is strongly recommended to carry out individual measurements in each area of investigation because the velocity used for the time–depth conversion directly influences the layer depths derived from the profiles.

Table 11.1. *Propagation velocities of GPR waves as derived from WARR measurements (own data)*

Location	Material	Velocity (m/ns)
Gravel pit (test)	Gravel	0.09
	Loose heap of gravel	0.16–0.18
Parzinn (Austrian Alps)	Loose debris (dolostone)	0.11–0.13
	Egesen moraine	0.09
Tegelberg (German Alps)	Debris, grass-covered (dolostone)	0.105
Dammkar (German Alps)	Loose debris (limestone)	0.10
Zugspitze (German Alps)	Loose debris (limestone)	0.12–0.128
Arnspitze (Austrian Alps)	Very loose, fresh debris (limestone)	0.14
Kühtai (Austrian Alps)	Loose debris (gneiss, mica-schist)	0.09–0.115
Turtmanntal (Swiss Alps)	Loose, coarse debris (gneiss, mica-schist)	0.12–0.14
	Egesen moraine	0.105
	LIA moraine	0.095
Cwm Cneifion	Loose debris (pyroclastic rocks)	0.14
(Snowdonia, Wales)	Grass-covered debris	0.10

Depending upon the coupling conditions to the ground surface, airborne reflections may pose a problem for radargram interpretation where the profile gets close to the rock face. Thus, it is important to note the distance to the rockwall and to protruding parts of it. Taking the propagation velocity in air (0.3 m/ns) as a basis, possible 'overhead' reflections can be identified in the radargram later.

The use of 50 MHz antennae turned out to be the best choice to achieve high penetration depth combined with satisfactory vertical resolution. The resolution of approximately 0.5 m (1/4 of the wavelength) was sufficient in most cases to detect the bedrock surface and major internal structures. A trace interval of 0.5 m proved sufficient to obtain detailed data. The application of 25 MHz antennae enhances the penetration depth and sometimes makes the prominent structures clearer; however, the accuracy is lower, which sometimes makes it harder to detect the bedrock surface or to interpret internal structures. Additional 100 MHz profiles turned out to be very helpful to detect internal stratification of the sediment layers, which facilitates the interpretation. However, due to the rough terrain (boulders) and the varying coupling conditions, 100 MHz profiles are sometimes marred by too noisy data.

In almost all of the measured profiles (more than 60 sections to date), the uppermost part of the talus bodies was characterised by a surface-parallel striation (see Figures 11.2 and 11.3 for examples). According to exposures, the stratification is probably due to a succession of stacked finer and coarser layers. There are many possible reasons for layering of slope deposits; e.g. DeWulf

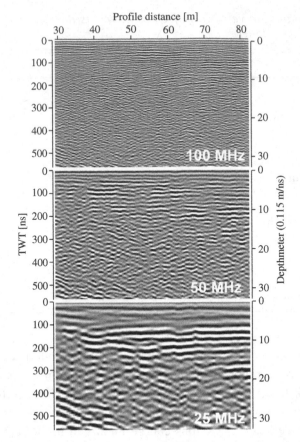

Figure 11.2.   Radargrams of 25, 50 and 100 MHz showing the same talus section in the Parzinn area. While the 100 MHz antennae are superior in depicting the talus stratification in the uppermost part, the 25 MHz section graphically shows the major subsurface layers (see Figure 11.4 for interpretation). The 50 MHz radargram provides a compromise between penetration and resolution.

(1988) or Sass and Krautblatter (2007) provide an overview. In the example (Figure 11.3), the layers extend in a straight line over almost the entire length of the talus, while the cross-profiles reveal slightly arched structures meeting each other discordantly. This leads to the three-dimensional image of cone-like accretion structures. There are different possible types of stratification discernable from scree slope radargrams, which can be linked to different processes involved in talus formation (Sass and Krautblatter, 2007).

The uppermost layer of 'striped' debris is frequently underlain by a zone of irregular reflection patterns (Figure 11.4). These reflections are probably due to glacial till or to coarse, unsorted rockfall deposits. A core of glacial sediments is very likely to occur in the steep cirques of the areas of investigation. However, under some talus bodies the 'irregular' layer showed very high electrical

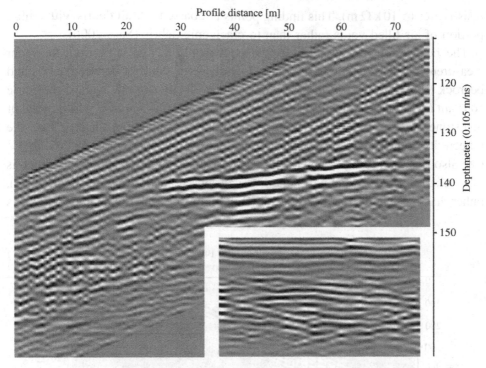

Figure 11.3. Topography-corrected longitudinal section (25 MHz) of the lower part of the 'Plenderlesee' talus cone in the Kühtai area. The surface-parallel stratification is clearly visible. The noticeable, almost horizontal reflector crossing the striation is due to the water table. A small lake is situated a few metres left of the profile's starting point. The small picture shows part of the cross-section. The reflectors are arched and meet each other discordantly.

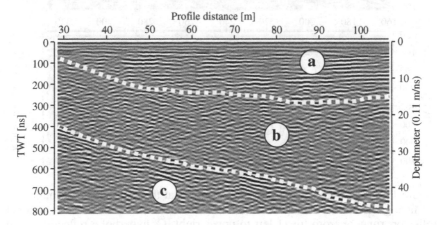

Figure 11.4. Typical sequence of stratified debris (a), irregular reflection patterns (b) and bedrock (c) from the Parzinn area. The irregular patterns are due to basal moraine or to coarse, unsorted rockfall deposits.

resistivities ($>10\,\mathrm{k}\Omega\,\mathrm{m}$). This finding points to coarse rockfall debris with a high portion of air-filled voids rather than to glacigenic debris (Sass 2006).

The bedrock surface is not always unequivocally recognisable from the radar measurements. This is due to the rather low dielectric contrast between debris and bedrock (see Table A4). From the propagation velocity (v), the average dielectric constant (ε) of talus sediments is estimated to be 4.6–9.0 ($v = c\varepsilon^{-0.5}$). Limestone or dolostone, for example, has $\varepsilon = 4$–8 according to literature data. Thus, the reflectivity of the talus/bedrock interface may range from approximately 20% to 0% (see also Equation (4.4), Chapter 4). As a consequence, the transition to bedrock is often characterised by fading of the internal talus reflections (A in Figure 11.5), rather than by a single, distinct reflector (B in Figure 11.5). If the bedrock surface is

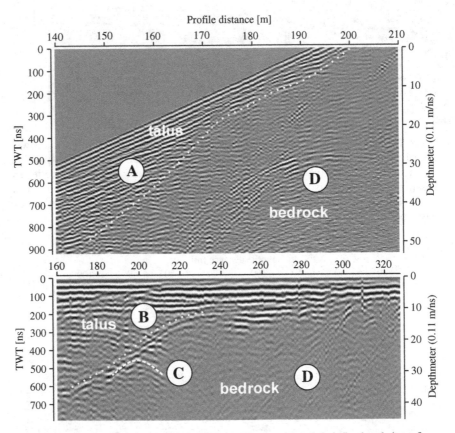

Figure 11.5. Different radar reflection patterns at the debris/bedrock interface. A: fading of the internal talus structures without distinct reflector, B: distinct reflector, running from lower left to upper right, C: hyperbolic reflection from buried edge of stratum, D: bedding planes of steeply dipping dolostone bedrock (weak lines dipping from upper right to lower left). Note that the mean reflectivity is much lower in the bedrock body than in the talus deposits.

characterised by steeply dipping, buried edges of strata, groups of reflection hyperbolae may occur in the radargrams, as typical for singular point-shaped reflectors (C in Figure 11.5). Additionally, bedding planes are frequently visible in the rock body (D in Figure 11.5), provided that the transmitted energy is high enough to propagate under the bedrock surface. In most cases, the overall reflectivity of bedrock is weaker than that of overlying sediment units. Assuming experience in radargram interpretation, the typical GPR patterns ('radar facies') of hard rock can usually be discriminated from those of loose sediments (Sass, 2007).

11.4 Conclusion

This contribution is a short draft showing the potential of GPR on loose sediments in alpine and arctic regions. However, universally valid instructions for measurement and interpretation cannot be provided. A cross-check of the results with boreholes, diggings or additional geophysical techniques is strongly advisable to avoid misinterpretation. This is particularly true when permafrost is present, as it may lead to ambiguous results when GPR is the only method applied (Otto and Sass 2006). An overview of coincidence and deviations in bedrock detection using different geophysical methods is provided in Sass (2007).

REFERENCES

DeWulf, Y. (1988). Stratified slope deposits. In *Advances in Periglacial Geomorphology*, ed. Clark, M. J., John Wiley & Sons, pp. 91–110.

Otto, J. C. and Sass, O. (2006). Comparing geophysical methods for talus slope investigations in the Turtmann valley (Swiss Alps). *Geomorphology*, **76**, 257–272.

Sass, O. (2006). Determination of the internal structure of alpine talus deposits using different geophysical methods (Lechtaler Alps, Austria). *Geomorphology*, **80**, 45–58.

Sass, O. (2007). Geophysical quantification of talus thickness and rockwall retreat in the eastern European Alps. *Journal of Applied Geophysics*, **6**, 254–269.

Sass, O. and Krautblatter, M. (2007). Debris-flow-dominated and rockfall-dominated scree slopes: genetic models derived from GPR measurements. *Geomorphology*, doi: 10. 1016/j.geomorph.2006.08.012.

12

GPR soundings of rock glaciers on Svalbard

I. Berthling, B. Etzelmüller, H. Farbrot, K. Isaksen,
M. Wåle and R. Ødegård

12.1 Introduction

GPR has been used for rock glacier investigations on the following four sites on
Svalbard: Hiorthfjellet rock glacier (78° 15' N, 15° 47' E) close to Longyearbyen;
four rock glaciers on the northwestern part of Prins Karls Forland (Forlandet)
(78° 50' N, 10° 30' E); Brøggerbreen rock glacier (78° 54' N, 11° 53' E) close to
Ny Ålesund; and on Nordenskiöldkysten (77° 53' N 13° 54' E). The Hiorthfjellet
rock glacier is a typical tongue-shaped, talus-derived rock glacier, which is
confined by the large bowl in the Hiorthfjellet mountainside. The Brøggerbreen
and Forlandet rock glaciers are lobe-shaped talus-derived rock glaciers situated at
the break of slope between the backing rockwall/talus slope and the valley bottom
(Brøggerbreen) or strandflat area (Forlandet). On Nordenskiöldkysten, a large,
complex talus-derived rock glacier, with a similar setting as those on Forlandet,
was investigated.

12.2 Methods

All GPR profiles were collected using antennae of 50 MHz centre frequency,
aligned perpendicular to the profile direction. Data from the Hiorthfjellet rock
glacier were obtained during the winter season in 1997 and followed the central
flowline of the Hiorthfjellet rock glacier. The length of the profile was 303 m. The
radar used was of the type pulseEKKO 100 (Sensors & Software Inc. Mississauga,
Canada). The transmitter/receiver antenna spacing was 2 m. Topographic data from
the rock glacier on Hiorthfjellet were obtained from a geodetic survey and a digital
altimeter. GPR profiles from Prins Karls Forland and Brøggerbreen rock glacier
were collected during summer 1998, using a RAMAC GPR system from Malå
Geoscience, Sweden. Data from Forlandet include the longitudinal profiles of four

Applied Geophysics in Periglacial Environments, eds. C. Hauck and C. Kneisel. Published by Cambridge
University Press. © Cambridge University Press 2008.

rock glaciers; all together nine transverse profiles and one CMP survey (see Chapter 4). The CMP survey yielded an electromagnetic velocity of 0.14 m/ns for the main part of the profile. This velocity was used in the processing of all other profiles. On Brøggerbreen rock glacier, one longitudinal profile was collected. Topographic data were obtained from a DEM on Forlandet and from a DGPS survey on Brøggerbreen rock glacier. The GPR from the Nordenskiöldkysten rock glacier crosses the main part of the rock glacier body, but does not extend up into the talus slope like the other longitudinal profiles do. The profiles were corrected for topography and processed using the software GRADIX (Interpex Ltd., Colorado, USA) on Forlandet, and REFLEXW (Sandmeier scientific software, Karlsruhe, Germany) on the Nordenskiöldkysten rock glacier.

12.3 Results and interpretation

12.3.1 Hiorthfjellet rock glacier (Isaksen et al. 2000 and Ødegård et al. 2003a, b)

The GPR profile from Hiorthfjellet is shown in Figure 12.1. The radar traces contain noise in the upper 2–3 m, caused by the air and surface waves and the block-rich surface. This part of the profile has been removed from Figure 12.1. The second zone extends to a depth of 15–20 m and contains a series of clear reflections. No detectable reflections are found below 25 m depth.

Starting at the rear of the rock glacier, the GPR record along the profile can be divided into four main sectors (Figure 12.1):

(i)   In the lower part of the talus slope, the upper 15 m of the profile from 285 to 300 m from the front, the reflectors are roughly parallel to the surface. This part of the profile is very short, but parallel reflections in the talus cone are expected and in accordance with results elsewhere.

(ii) About 285 m from the front, at the break of slope between talus and rock glacier, the reflectors dip down into the rock glacier.

(iii) Between 110 and 260 m from the front, clear and continuous reflections can be observed. They rise and slant up towards the surface slope.

(iv) In the frontal 90 m of the rock glacier, the reflectors are slightly more parallel to the surface.

The radar data give no indication of the thickness of the rock glacier, or any shear horizon in the upper 20 m of the rock glacier. The layers are interpreted as deposits from major rock fall events that may have covered seasonal or avalanche snow in the rooting zone of the rock glacier. From the GPR profile it is apparent that these layers experience rotation and flexure in the upper part of the rock glacier as layers first dip down and then up again relative to the rock glacier surface. A dip down may be caused by repeated accumulation of debris and ice within a narrow area. The rapid

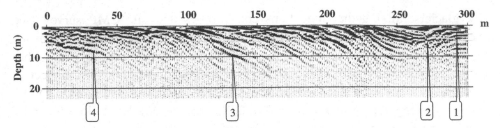

Figure 12.1. The GPR profile from the Hiorthfjellet rock glacier. The profile starts at the front of the rock glacier and ends at the lower part of the talus cone. The horizontal length of the reflection horizons has an extent of approximately 30–100 m. Callouts refer to features described in the text.

change in reflector angle between 250 and 300 m from the front may be caused by a bedrock threshold beneath this rock glacier (Berthling *et al.* 2000).

12.3.2 Forlandet rock glaciers (Berthling et al. 2000, 2003)

Maximum depth penetration was about 30 m. The longitudinal profiles revealed a system of reflections that was relatively consistent from one rock glacier to the other. The reflectors are commonly a few tens of metres long. Also the cross profiles on the Forlandet rock glaciers had similarities. The following refer to Figure 12.2, which includes the longitudinal profile and three cross profiles from rock glacier 9 on Forlandet (Berthling *et al.* 2003):

In the upper part of the talus cone, the reflections are parallel to the surface (L1). This feature is better developed on rock glaciers 12 and 15 (Berthling *et al.* 2000) than on the present example. Towards the transition zone to the rock glacier, these reflectors start to slant against the surface slope (L2). This inclination of the layers prevails towards the front, although the layers are mainly more irregular in the frontal zone (L3). Rock glaciers 7, 9 and 12 all have a sector about 5 m deep beneath the surface reflectors where no or only very faint reflectors are found (L4). This is interpreted as a layer of more or less clean ice. It may, however, be partly an artefact caused by the AGC gain applied to the data (see Chapter 4). Further important features of this particular longitudinal profile are (L5) a prominent reflector on rock glacier number 9, interpreted as the rock glacier–bedrock interface and (L6) linear reflectors rising upslope, originating from above-surface reflections from the mountain backwalls (see also Chapter 11).

The transverse profiles can be divided into four zones. First (T1), where there are sharp and continuous reflectors in the active layer. Then (T2), a zone corresponding to (L5), is found beneath the active layer. Here no or only few vague reflectors are found. Further down, scattered dipping or wavy reflectors are found (T3). They may be well developed, but seldom along any great distance. They are

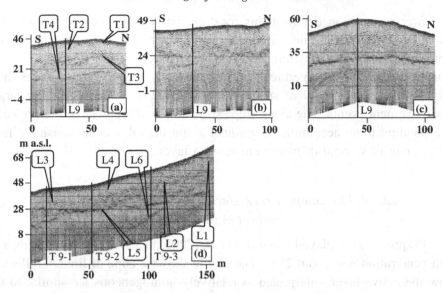

Figure 12.2. GPR longitudinal and cross profiles (T 9-1 (a), T 9-2 (b) and T 9-3 (c)) from rock glacier 9 on Forlandet. The longitudinal profile (d) starts at the front of the rock glacier and ends high up on the talus cone. The location of the cross profiles (T 9-1 to T 9-3) are displayed on the longitudinal profile, and the crossing of the longitudinal profile (L 9) is similarly shown on each of the cross profiles. All cross profiles are plotted looking towards the front (N: north and S: south). Callouts refer to specific features described in the text.

mainly sub-parallel to the surface slope, especially on or near the talus cones, but opposite examples are found. The depths of these reflectors often match well with depths to reflectors on the longitudinal profile. Finally, a set of reflectors or a sharp reflector is found, developed more or less continuously across the profile (T4), which is quite probably the bedrock interface (Figure 12.2c). Some of these reflectors may possibly be interpreted as a shear zone. The effect of the AGC gain is seen below the bedrock reflector in Figure 12.2c and along parts of this same reflector in Figure 12.2d.

12.3.3 Brøggerbreen rock glacier (Berthling et al. 2000)

The GPR profile from Brøggerbreen is displayed in Berthling *et al.* (2000) and is not repeated here. The longitudinal profile shows reflectors comparable to those found on Forlandet. However, it lacks irregular reflectors in the frontal part and the reflectors are generally found closer to the surface. Further, some reflectors at the extreme lower end of the talus cone dip down into the rock glacier when looking downslope. This is somewhat similar to what was found on the Hiorthfjellet rock glacier. Beneath a depth of about 10 m, no reflectors are found.

The lower penetration depth at this rock glacier may be caused by a higher content of fine material and ions in the pore water, due to weathering of the dolomite-derived surface material.

On Forlandet and Brøggerbreen rock glaciers, the layering structure is most likely developed by burial of snow or ice-supersaturated permafrost by larger mass movement events. The along-slope structural development is interpreted as corresponding to an accumulation gradient along the talus cone, causing differences in rate of vertical displacement along a layer.

12.3.4 The complex rock glacier on Nordenskiöldkysten (Farbrot et al. 2005)

The GPR profile is displayed in Farbrot *et al.* (2005) and is not repeated here. The depth penetration was about 25 m. The profile shows a zone of limited reflectors below the active layer, interpreted as relatively homogeneous ice similar to the rock glaciers on Forlandet. This zone disappears towards the front of the rock glacier. Further down into this rock glacier, a zone of scattered reflectors is found. These are wavy and dipping in an apparently chaotic pattern. Also a few continuous reflectors occur. None of these features have any close resemblance to features found on the GPR profiles described above.

12.4 Discussion

The longitudinal profiles from all Svalbard rock glaciers show some common development. Similarities are best developed between Hiorthfjellet, Forlandet and Brøggerbreen rock glaciers, where the along-slope structural development is fairly comparable. This indicates a similar development with respect to accumulation processes and internal deformation. The Nordenskiöldkysten and Forlandet rock glaciers share a layer of few reflections beneath the active layer, while the Nordenskiöldkysten rock glacier lacks the along-slope structural development. This may be caused by a more complex history of development, which also is evident from the more chaotic surface topography, with several prominent ridges and compressions on this rock glacier. It may be that it resembles the more chaotic reflectors towards the front found on Prins Karls Forland. However, as the radar profile on the Nordenskiöldkysten rock glacier does not extend up into the talus slope, a real comparison with the other longitudinal profiles is difficult.

The GPR profiles generally reveal structural features that are very important for developing conceptual models for the development of these high arctic rock glaciers. On smaller rock glaciers, the bedrock interface was partly found and measurements with 25 MHz antennae could possibly give even better control on the vertical dimension of the rock glaciers. On some of the rock glaciers a reflector that might correspond to a possible shear zone was found. However, without a borehole to confirm the location of a shear zone, such interpretations are probably of little scientific value as they represent adaptions of data to present models rather than data to test these models.

REFERENCES

Berthling, I., Etzelmüller, B., Isaksen, K. and Sollid, J.L. (2000). The rock glaciers on Prins Karls Forland (II): GPR soundings and the development of internal structures. *Permafrost and Periglacial Processes*, **11**, 357–369.

Berthling, I., Etzelmüller, B., Wåle, M. and Sollid, J.L. (2003). Use of ground penetrating radar (GPR) soundings for investigating internal structures in rock glaciers. Examples from Prins Karls Forland, Svalbard. *Zeitschrift für Geomorphologie, Supplement*, **132**, 103–121.

Farbrot, H., Isaksen, K., Eiken, T., Kääb, A. and Sollid, J.L. (2005). Composition and internal structures of a complex rock glacier at Nordenskiöldkysten, Svalbard. *Norwegian Journal of Geography*, **59**, 139–148.

Isaksen, K., Ødegård, R.S., Eiken, T. and Sollid, J.L. (2000). Composition, flow and development of two tongue-shaped rock glaciers in the permafrost of Svalbard. *Permafrost and Periglacial Processes*, **11**, 241–257.

Ødegård, R.S., Isaksen, K., Eiken, T. and Sollid, J.L. (2003a). A conceptual model of Hiorthfjellet rock glacier, Svalbard. *Proceedings of the 8th International Conference on Permafrost*, Zürich, Switzerland, 839–844.

Ødegård, R.S., Isaksen, K., Eiken, T. and Sollid, J.L. (2003b). Terrain analyses and surface velocity measurements of the Hiorthfjellet rock glacier, Svalbard. *Permafrost and Periglacial Processes*, **14**, 359–365.

13

Arctic glaciers and ground-penetrating radar. Case study: Stagnation Glacier, Bylot Island, Canada

T. Irvine-Fynn and B. Moorman

13.1 Introduction

This case study presents results from multiple ground-penetrating radar (GPR) surveys conducted over a single ablation season at a polythermal glacier in the Canadian Arctic. Recent advances in both equipment functionality and data analysis have allowed researchers to examine notions of spatial variation in subsurface conditions, yet to date the full potential of the geophysical tool has not been exploited. A total of 30 km of radar profile data were collected illustrating how seasonal hydrothermal development can be observed in glaciers, including the appearance and disappearance of hydrological features. However, this research exemplified issues that should be borne in mind when undertaking glacier-based GPR, particularly focusing on resolution, interpolation, increasing noise and short-term temporal variability in glacier ice conditions.

Over the past 50 years, radio-echo sounding (RES) and GPR have been increasingly employed on glaciers to determine a number of geometric and structural conditions. Reflection of a proportion of the impulse wave occurs where there is an abrupt, subsurface transition in dielectric constant (κ). The stark contrast between air ($\kappa = 1$), water ($\kappa = 80$), sediment ($\kappa \approx 25$) and ice ($\kappa \approx 3.5$) enables the reconstruction of subsurface structures from geophysical surveys. RES was initially focused upon the determination of ice thickness and thus bed topography for ice sheets (e.g. Evans 1963, Robin *et al.* 1969, Davis *et al.* 1973, Bentley *et al.* 1979, Hodge *et al.* 1990). More recently, RES applications began to examine the thermal regime of high-latitude glaciers, using the dielectric difference between cold (water-poor) and temperate (water-rich) ice (e.g. Hagen and Sætrang 1991; Björnsson *et al.* 1996).

The scattering of the RES pulse in thermally and/or structurally complex glacier ice leads to difficulties in identifying unambiguous returns (e.g. Dowdeswell *et al.*

Applied Geophysics in Periglacial Environments, eds. C. Hauck and C. Kneisel. Published by Cambridge University Press. © Cambridge University Press 2008.

1984, Kotlyakov and Macheret 1987). And, commonly, RES would be conducted using an airborne radar system, which limits the penetration of the impulse wave. Gaining data pertaining to englacial structures, particularly hydrological and sedimentological features, was problematic using RES systems.

As a result of the limitations RES had for shallow subsurface (glaciological) applications, GPR was developed to address issues of more detailed resolution. Being a specific form of RES, GPR uses a short-pulse, 10–1000 MHz impulse wave. Higher-frequency GPR systems have greater vertical resolution (200 MHz can resolve ~0.5 m features in ice) but have limited penetration depth. Use of lower-frequency antennae allows features to be observed over greater depths, but has lessened vertical resolution (50 MHz has theoretical vertical resolution of ~1 m in ice). Significantly, features smaller than the vertical resolution can be detected using GPR, but their dimensions are irresolvable. However, although feature dimensions may or may not be resolved, the dielectric interface causing the radar pulse reflection generates a three-half-cycle reflection wavelet (Arcone *et al.* 1995). It is therefore possible to use reflection polarity to interpret the nature of unknown subsurface reflectors.

In glaciology, higher-frequency GPR has been used to examine mass balance and map previous ice surfaces (e.g. Kohler *et al.* 1997, Pälli *et al.* 2002, 2003). Recent work focusing on radar pulse reflection characteristics includes identification of sediment structures (e.g. Murray *et al.* 1997) and englacial water and air voids (e.g. Walford *et al.* 1986, Kennett 1989, Arcone *et al.* 1995, Moorman and Michel 2000, Stuart *et al.* 2003). Researchers examining spatial variability have focused on velocity and backscatter analysis to estimate changes in interstitial water content (e.g. Macheret *et al.* 1993, Murray *et al.* 2000, Pettersson *et al.* 2004). However, few GPR applications have examined temporal change. Early work by Jacobel and Raymond (1984) compared radar signatures at Variegated Glacier, Alaska, and suggested that temporal variation indicated changes in internal water pressure, while Pettersson *et al.* (2003) compared GPR surveys from 1989 and 2001 to examine cold surface layer thinning.

This case study documents GPR surveys conducted over the course of the 2002 ablation season at polythermal Stagnation Glacier, principally investigating whether seasonal hydrothermal changes can be observed by using a geophysical method.

13.2 Field site

Stagnation Glacier is located on the southeastern edge of the Byam Martin ice field, Bylot Island (Figure 13.1). The glacier is about 14 km², ranging from 320 m a.s.l. to approximately 1650 m a.s.l. GPR surveys conducted by Moorman and

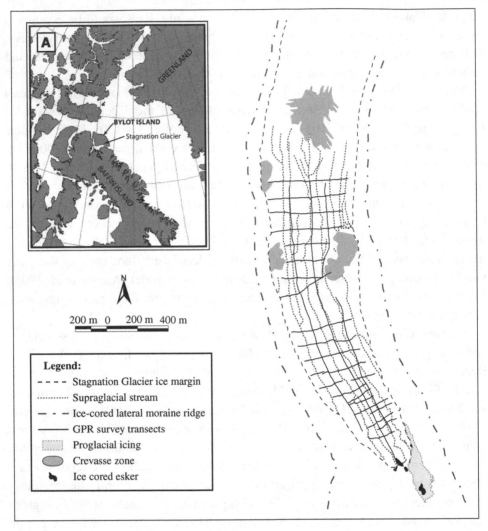

Figure 13.1.   Map of Stagnation Glacier's lower ablation zone. GPR survey
transects conducted during summer 2002 are marked. Inset A shows location of
Bylot Island.

Michel (2000) indicate that the glacier is polythermal. The study area included
the lower 2 km of the ablation area, a zone effectively devoid of moulins, displaying negligible surface debris and with small 'crevasse zones' close to the
glacier margins. The surface topography is dictated by deeply incised supraglacial stream valleys, particularly over the lower 800 m. Emergent esker forms at
the contemporary glacier snout indicate englacial hydrological architecture may
dominate Stagnation Glacier's hydrology.

13.3 Field methods

A Sensors & Software Inc. pulseEKKO 100 system was used for the GPR surveys during 2002. Profiles were obtained using bistatic antennae in a parallel broadside configuration. To maximise the depth of penetration, 50 MHz antennae were used with a 1000 V transmitter. The antennae were dragged as a sled approximately 10 m ahead of the control unit (Figure 13.2). Antenna separation was defined by the sled configuration at 1.55 m. This distance is sub-optimal for 50 MHz antennae; however, the problem of reverberations caused by small antenna separation was limited to the near surface, which in glaciological terms is unimportant for delineation of englacial hydrological features. Along the profiles, traces were collected at a set spacing using an odometer wheel.

It was assumed that minimal englacial hydrological change took place during the course of each survey, as no significant hydrological events occurred (e.g. rain storms, proglacial upwelling). However, to ensure survey repeatability, a grid survey scheme was set up. The grid was marked out using forestry survey string with flagging tape markers to guarantee visibility. The grid was temporarily anchored to the glacier surface with supraglacial rocks. The greater number of cross-glacier profiles were assumed to cross-cut glacial drainage features, which

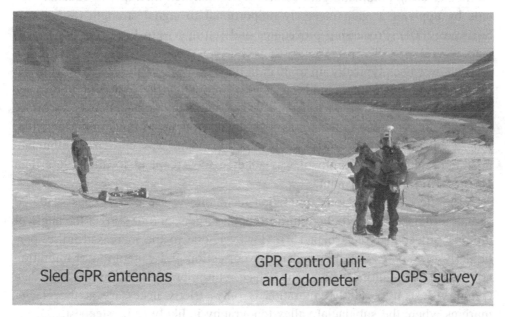

Figure 13.2. Illustration of GPR surveying set-up. A handheld odometer was used to trigger traces from the control unit as the antennae in a sledge configuration were pulled over the glacier surface. GPS data were used to topographically correct GPR profiles.

would be expected to be oriented sub-parallel to the glacier's long axis. Three complete GPR surveys were begun on Julian Day (JD) 175, JD200 and JD213. Each survey following the marked grid took up to 8 days to complete, and used identical settings for data acquisition.

A Trimble differential global positioning system (DGPS) was used to collect positional data throughout the study site. DGPS and GPR data were collected simultaneously along profiles during the first and third surveys (see Figure 13.2). Full details of DGPS data accuracy and precision at Stagnation Glacier can be found in Walter (2003).

13.4 Processing methods

Due to the proximity of transmitting and receiving antennae, a low-frequency 'wow' may be induced superimposed on each trace; DEWOW, being a time filter, reduces this unwanted signal saturation. Use of additional data-enhancement procedures was kept to a minimum for ease of comparison between profiles. A three-point down-trace averaging procedure was applied to aid reduction of undesirable, high-frequency noise. A gain function is usually applied to radar data to emphasise weaker returns. Initial interpretation of the GPR data was facilitated using automatic gain control (AGC), which attempts to equalise signals by applying a gain inversely proportional to signal strength (Sensors & Software 2001). Processing procedures and settings were kept constant for all data.

With crevasses typically up to 30 m in depth, and possibly deeper in cold ice (Summerfield 1991, Benn and Evans 1998), following the closure of a crevasse, air, sediment or water filled voids may exist englacially at such depths. Incorrect identification of these reflectors as hydrological englacial conduits is possible, thus a cut-off depth of 400 ns, equivalent to 34 m, was chosen to eliminate misinterpretation. This protocol was adhered to, except at locations less than 100 m from the glacier snout.

A simple one-dimensional migration technique was used to focus scattered signals, collapsing hyperbolic returns to point targets, and to correct the angle of dipping reflections (see Chapter 4). However, due to the increased noise, particularly in Surveys 2 and 3, under/over-correction and generation of migration artefacts made identification of subsurface returns problematic in migrated data (Figure 13.3). Importantly, migration is only strictly necessary at the glacier margins where the subglacial valley topography is likely to be steepest.

Static topographic correction of the GPR data was achieved with the DGPS survey. Positional errors and uncertainty were assumed to be negligible given that the geometrical spreading of the impulse wave from 50 MHz antennae results in

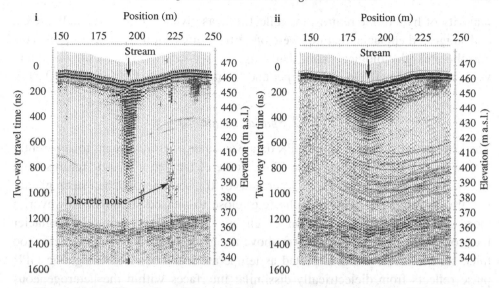

Figure 13.3. Illustration of the problems encountered in migration of GPR data collected at Stagnation Glacier. (i) The raw unmigrated profile, and (ii) the migrated data. Note the problems of over-migration in locations crossing supraglacial streams. The up-turned asymptotes and loss of coherent reflection at the bed (approximately 365 m a.s.l.) is typical of a low-velocity (water) surface (see Arcone *et al.* 1995). Also note the artefacts generated by near-surface and discrete noise, potentially concealing focused point-source hyperbolae.

footprint dimensions of 28.4 m at 40 m depth and 55.8 m at 80 m depth. Uncertainty in bed reflector depth was less than 5 m, 92% of the tie points were within this error bracket.

13.5 Results

The complete 2002 GPR data from Stagnation Glacier are presented in Irvine-Fynn (2004). However, several key results will be mentioned here.

13.5.1 General

Reflections seen in the GPR data were categorised as four key types (Figure 13.4):

(i)   dense chaotic returns within the ice body,
(ii) dipping slopes (indicative of the glacier bed),
(iii) continuous linear or quasi-linear non-bed reflectors,
(iv) hyperbolic returns from point sources.

Despite the theoretical resolution of 0.85 m, close inspection of GPR data indicated that true vertical resolution was of the order of 1.2 m. For the vast majority of hyperbolic returns, the reflection was given as only three half cycles, and suggested englacial features were of vertical dimensions less than 1.2 m. This corroborates glaciological work that suggests, typically, englacial channels and voids are less than 0.5 m (e.g. Harper and Humphrey 1995, Copland *et al.* 1997).

13.5.2 Spatial variations

The key spatial variation in the GPR data was changes from 'clean' to apparently 'noisy' trace sections (i on Figure 13.4). Some of the dense chaotic returns were seen as either off-line reflections where survey transects approached steep glacier margins or supraglacial topography (e.g. close to incised streams). However, a more spatially distributed region of chaotic returns was found in upglacier locations, ranging from 0 m to 60 m above the glacier bed. This region, being too thick for basal ice, was interpreted as temperate ice. The electromagnetic GPR pulse reflects from dielectrically dissimilar interfaces within the heterogeneous ice matrix which are separated by distances significantly less than radar resolution (i.e. < 0.85 m) thereby generating chaotic, noisy returns. Measurements of

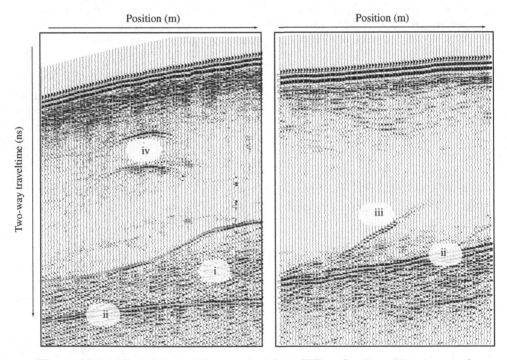

Figure 13.4.   Illustration of four typical key GPR reflections: (i) a zone of chaotic returns occurring above the clearly defined bed reflection; (ii) the dipping bed slope; (iii) a linear non-bed reflection; and (iv) hyperbola created by point-source reflector at the apex of the curve.

ice temperature would be required to confirm this; however, previous researchers have indicated that the contrast between 'clean' and 'noisy' trace sections correlates to the change from cold to temperate ice (e.g. Ødegård *et al.* 1992, Björnsson *et al.* 1996, Ødegård *et al.* 1997, Pettersson *et al.* 2003).

Interestingly, the temperate zone also displayed a variation itself, with a less noisy region approximately following the glacier centre line (see Irvine-Fynn *et al.* 2006). This suggests, as has been implied in previous work, that even within temperate ice there is a spatial variation in water content (e.g. Pettersson *et al.* 2004).

13.5.3 Temporal variations

The most crucial element of this case study was the observation of a number of temporal variations during the course of the ablation season. The first was the change in spatial extent of the temperate ice. Between Survey 1 and Survey 3, the temperate ice zone appeared to retreat upglacier by approximately 200 m. This corresponded to a decrease in interstitial water content of some $17\,000\,\text{m}^3$ (Irvine-Fynn *et al.* 2006).

Another dynamic that was observed was the appearance and disappearance of a number of features interpreted as both hydrological and air voids (Figure 13.5). The implication of these changes in radar signature suggests temporary englacial storage of water, the drainage of small conduits, or the elimination of voids within the ice body.

The last temporal variation was an apparent increase in down-trace noise and signal attenuation between the three glacier-wide surveys. The apparent signal attenuation may be linked to the change in hydrothermal condition of the glacier surface. The presence of slush zones, particularly early in the season proximate to supraglacial streams, provided strong returns at the beginning of traces. However, as the season progressed, despite the disappearance of slush, the surface ice heterogeneity should have increased due to changes in surface and near-surface ice temperature as the ablation progressed. This could potentially attenuate the radar pulse without any other significant effect on the data. Despite attenuation, penetration depths of 120 m were still readily achievable.

Down-trace noise may also be an indication of the loss of homogeneity in the surface and near-surface ice. Kotlyakov and Macheret (1987) suggested that scattering of electromagnetic pulses may be caused where enhanced melting and/or refreezing during the ablation season leads to increasing sub-resolution englacial heterogeneities, potentially within the temperate ice zone in addition to the surface and near-surface. Some of the noise noted was of relatively high amplitude and more distinctive. One possible cause of this noise could have been cracks in the optic fibre cables caused by use in low temperatures.

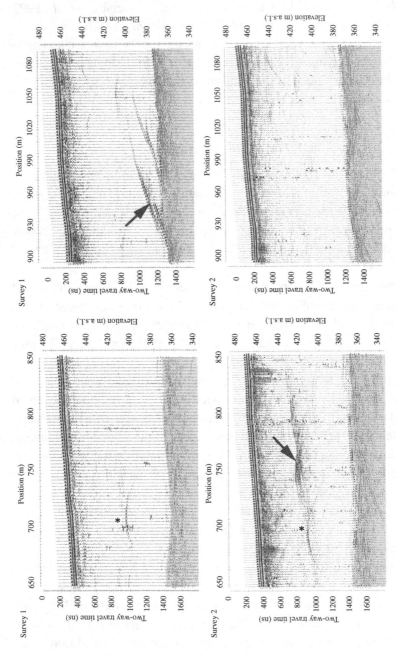

Figure 13.5. Comparison of GPR data between Survey 1(top) and Survey 2 (bottom) indicating temporal change along profiles as highlighted by bold arrows. Top: Note quasi-linear reflections dipping downglacier between 910 m and 1050 m. These reflections were absent from the profiles from both Survey 2 and Survey 3. Bottom: Note the appearance of the hyperbola and noise sequence between 740 m and 800 m. These reflections were only present in Survey 2. Also shown is a persistent hyperbola at 420 m a.s.l. (marked by the *), the slight positional change is an artefact of the minor change in profile-line location and trace-spacing adjustment. Importantly, between surveys, the position of the bed remains consistent, indicating good agreement between successive surveys.

13.6 Discussion

This case study clearly demonstrates the potential GPR has for the investigation of the internal hydrological dynamism of glaciers.

The ability to observe the appearance and disappearance of hydrological features means GPR is a method by which changes in englacial water routing may be observed. To date, dye tracing and geometric drainage modelling have been the principal methods employed to reconstruct a glacier's internal drainage system. However, as has been demonstrated, GPR provides an alternative method and one which rather than providing theoretical models is able to 'directly' observe englacial conduit locations.

Using GPR to identify a change in the interstitial water content of a polythermal glacier raises a number of significant questions. Firstly, demarcating the upper surface of a temperate ice zone is fraught with difficulties. This study reaffirms that a more sophisticated GPR configuration may be required to identify the zero-degree isotherm. While some research suggests the transition from noisy to clear trace correlates well with ice temperatures, an implication of interstitial drainage is whether GPR can be used to map thermal regimes with confidence. If the glacier interior represents a water reservoir, it begs the question of how this region is 'recharged' during winter months. Further, does the drainage of interstitial water have a reciprocal impact on the thermal condition of an arctic glacier and, if interstitial water drains, do polythermal glaciers exhibit a hydrostatically low pressure interior? Thus, crucially, the time at which a GPR survey is conducted over a glacier may or may not accurately reveal the glacier's thermal condition.

However, it is important to make note of issues that have to be borne in mind. The first is the problem of interpolation, particularly in efforts to reconstruct drainage routes. The distances between survey transects were typically 50–100 m, which were sufficient to make interpolation between transects problematic. For researchers wishing to map specific englacial structures it is important to use higher-density profiling (e.g. Moorman and Michel 2000).

While some hyperbolic reflections may occur close to the bed interface, determination of the presence of water versus the simple but transitional change from ice to basal ice or bed sediments is problematic. This is emphasised by the fact that the polarity of reflections would be identical. Therefore, locating sub-glacial channels remains challenging. One manner in which this might be addressed is the use of bed reflection power to give an indication of the likelihood of water presence (e.g. Gades *et al.* 2000).

The choice of antennae dictates the resolution available. To resolve englacial conduits, this study suggests higher-frequency antennae are required. It is therefore important that work continues to advance the capabilities of GPR systems.

13.7 Conclusions

Despite a number of limitations, GPR provides a successful method by which to examine glacial hydrological and thermal characteristics. Changes in a glacier's hydrothermal condition and the location of hydrological features such as englacial conduits can potentially be mapped using a geophysical method. In this respect GPR acts as a technique able to address issues that have so far eluded glacial hydrologists.

However, this case study suggests that glaciers are temporally variable in hydrological condition. We suggest that frequent, repeated GPR studies are a good way to examine the seasonal progression of the thermal and hydrological system. Such sequential GPR surveys might prove to complement dye tracing as a method by which to examine the location, routing and dynamism of glaciers' hydrological architecture.

REFERENCES

Arcone, S. A., Lawson, D. E. and Delaney, A. J. (1995). Short-pulse radar wavelet recovery and resolution of dielectric constants in englacial and basal ice in Matanuska Glacier, Alaska, *Journal of Glaciology*, **41**, 68–86.

Benn, D. I. and Evans, D. J. A. (1998). *Glacier and Glaciation*. Arnold.

Bentley, C. R., Clough, J. W., Jezek K. C. and Shabtaie, S. (1979). Ice thickness patterns and the dynamics of the Ross Ice Shelf, Antarctica. *Journal of Glaciology*, **24**, 287–294.

Björnsson, H., Gjessing, Y., Hamran, S.-E., Hagen, J., Liestøl, O., Pálsson, F. and Erlingsson, B. (1996). The thermal regime of sub-polar glaciers mapped by multi-frequency radio-echo sounding. *Journal of Glaciology*, **42**, 23–32.

Copland, L., Harbor, J., Gordon, S. and Sharp, M. (1997). The use of borehole video in investigating the hydrology of a temperate glacier. *Hydrological Processes*, **11**, 211–224.

Davis, J. L., Halliday J. S. and Miller, K. J. (1973). Radio echo sounding on a valley glacier in east Greenland. *Journal of Glaciology*, **12**, 87–91.

Dowdeswell, J. A., Drewry, D. J., Liestøl, O. and Orheim, O. (1984). Radio echo-sounding of Spitsbergen glaciers: problems in the interpretation of layer and bottom returns. *Journal of Glaciology*, **30**, 16–21.

Evans, S. (1963). Radio-echo techniques for the measurement of ice thickness. *Polar Record*, **11**, 406–410.

Gades, A. M., Raymond, C. F., Conway, H. and Jacobel, R. W. (2000). Bed properties of Siple Dome and adjacent ice streams, West Antarctica, inferred from radio-echo sounding measurements. *Journal of Glaciology*, **46**, 88–94.

Hagen, J. O. and Sætrang, A. (1991). Radio-echo soundings of sub-polar glaciers with low frequency radar. *Polar Research*, **9**, 99–107.

Harper, J. T. and Humphrey, N. F. (1995). Borehole video analysis of a temperate glacier's englacial and subglacial structure: implications for glacier flow models. *Geology*, **23**, 901–904.

Hodge, S. M., Wright, D. L., Bradley, J. A., Jacobel, R. W., Skou, N. and
Vaughn, B. (1990). Determination of the surface and bed topography in Central
Greenland. *Journal of Glaciology*, **36**, 17–30.

Irvine-Fynn, T. D. L. (2004). *A Non-Invasive Investigation of Polythermal
Glacial Hydrology: Stagnation Glacier, Bylot Island, Nunavut, Canada*. M.Sc.
Thesis, University of Calgary, Alberta, 365pp.

Irvine-Fynn, T. D. L., Moorman, B. J., Williams, J. L. M and Walter, F. S. A. (2006).
Seasonal changes in ground-penetrating radar signature observed at a polythermal
glacier, Bylot Island, Canada. *Earth Surface Processes and Landforms*, **31**(7),
892–909.

Jacobel, R. and Raymond, C. (1984). Radio echo-sounding studies of englacial
water movement in Variegated Glacier, Alaska. *Journal of Glaciology*, **30**,
22–29.

Kennett, M. I. (1989). A possible radio-echo method of locating englacial and subglacial
waterways. *Annals of Glaciology*, **13**, 135–139.

Kohler, J., Moore, J., Kennett, M., Engeset, R. and Elvehøy, H. (1997). Using
ground-penetrating radar to image previous years' summer surfaces for mass
balance measurements. *Annals of Glaciology*, **24**, 355–360.

Kotlyakov, V. M. and Macheret, Y. Y. (1987). Radio echo-sounding of sub-polar glaciers
in Svalbard: some problems and results of Soviet studies. *Journal of Glaciology*,
9, 151–159.

Macheret, Y. Y., Moskalevsky, M. Y. and Vasilenko, E. V. (1993). Velocity of
radio waves in glaciers as an indicator of their hydrothermal state, structure and
regime. *Journal of Glaciology* **39**, 373–384.

Moorman, B. J. and Michel, F. A. (2000). Glacial hydrological system characterization
using ground penetrating radar. *Hydrological Processes*, **14**, 2645–2667.

Murray, T., Gooch, D. L. and Stuart, G. W. (1997). Structures within the surge front at
Bakaninbreen, Svalbard, using ground penetrating radar. *Annals of Glaciology*, **24**,
122–129.

Murray, T., Stuart, G. W., Fry, M., Gamble, N. H. and Crabtree, M. D. (2000).
Englacial water distribution in a temperate glacier from surface and borehole radar
velocity analysis. *Journal of Glaciology*, **46**, 389–398.

Ødegård, R. S., Hamran, S.-E., Bø, P., Etzelmüller, B., Vatne, G. and Sollid, J. L. (1992).
Thermal regime of a valley glacier, Erikbreen, northern Spitsbergen. *Polar
Research*, **11**, 69–79.

Ødegård, R. S., Hagen, J. O. and Hamran, S.-E. (1997). Comparison of radio-echo
sounding (30–1000 MHz) and high resolution borehole-temperature measurements
at Finsterwalderbreen, southern Spitsbergen, Svalbard. *Annals of Glaciology*, **24**,
262–267.

Pälli, A., Kohler, J., Isaksson, E., Moore, J. C., Pinglot, J. F., Pohjola, V. A.
and Samuelsson, H. (2002). Spatial and temporal variability of snow accumulation
using ground-penetrating radar and ice cores on a Svalbard glacier. *Journal of
Glaciology*, **48**, 417–424.

Pälli, A., Moore, J. C. and Rolstad, C. (2003). Firn-ice transition-zone features of four
polythermal glaciers in Svalbard seen by ground-penetrating radar. *Annals of
Glaciology*, **37**, 298–304.

Pettersson, R., Jansson, P. and Holmlund, P. (2003). Cold surface layer thinning
on Storglaciären, Sweden, observed by repeated ground penetrating radar
surveys. *Journal of Geophysical Research*, **108**, F06004, doi:10.1029/
2003JF000024.

Pettersson, R., Jansson, P. and Blatter, H. (2004). Spatial variability in water content at the cold-temperate transition surface of the polythermal Storglaciären, Sweden, *Journal of Geophysical Research*, **109**, F02009, doi:10.1029/2003JF000110.

Robin, G. deQ., Evans, S. and Bailey, J. T. (1969). Interpretation of radio-echo sounding in polar ice sheets. *Philosophical Transactions of the Royal Society of London, Series A*, **265**, 437–505.

Sensors & Software (2001). *Win_EKKO: User's guide, version 1.0.* Sensors & Software Inc., Mississauga, Ontario, 77pp.

Stuart, G., Murray, T., Gamble, N., Hayes, K. and Hodson, A. (2003). Characterization of englacial channels by ground-penetrating radar: An example from Austre Brøggerbreen, Svalbard. *Journal of Geophysical Research*, **108**, B2535, doi:10.1029/2003JB002435.

Summerfield, M. A. (1991). *Global Geomorphology.* Longman.

Walford, M. E., Kennett, M. I. and Holmlund, P. (1986). Interpretation of radio echoes from Storglaciären, northern Sweden. *Journal of Glaciology*, **32**, 39–49.

Walter, F. S. A. (2003). *Glacier Fluctuations and Glacier Hypsometry on Bylot Island, Nunavut, Canada, Using Remote Sensing Methodology.* M.Sc. thesis, University of Calgary, Alberta, 159pp.

14

Mapping of subglacial topography using GPR for determining subglacial hydraulic conditions

K. Melvold and T.V. Schuler

14.1 Introduction

Recently initiated coal mining in Svea Nord, Svalbard, by Store Norsk Spitsbergen Kullkompani (SNSK), has encountered problems during summer seasons due to water intrusions that originate from the glacier above the mine (Melvold *et al.* 2003, Schuler *et al.* 2005). The coal mine Svea Nord is situated in bedrock 150–350 m below the bed of the Gruvefonna ice cap (Figures 14.1 and 14.2a). In Svalbard, ice-free land areas have continuous permafrost with thickness varying from less than 100 m near sea-level up to 500 m in the higher mountains (Liestøl 1977). However, beneath the accumulation areas of larger glaciers and ice caps, taliks (unfrozen areas within an otherwise frozen region) may exist (Liestøl 1977). Mining in permafrost regions, where it may encounter taliks, is a known challenge (Tolstikhin and Tolstikhin 1976), since, as in this case, a permeable talik enables the leakage of water into the mine. To stop the leakage from the usually continuous hydraulic system of the subpermafrost aquifer can be difficult, or even impossible. After removal of the coal, the roof of the mine is brought down to reduce the load on the remaining coal layer, thereby presumably increasing the permeability of the rock material between the glacier bed and the mine.

During summers 2003 and 2004, the water inflow in Svea Nord mine reached maximum rates of 2000 and 5000 $m^3 h^{-1}$, respectively. As it is intended to enlarge the mine significantly, predictions of the total volume and the rate of future water intrusions are important design and planning requirements. The planned extent of the mine underneath Gruvefonna is outlined in Figure 14.1. It is

Applied Geophysics in Periglacial Environments, eds. C. Hauck and C. Kneisel. Published by Cambridge University Press. © Cambridge University Press 2008.

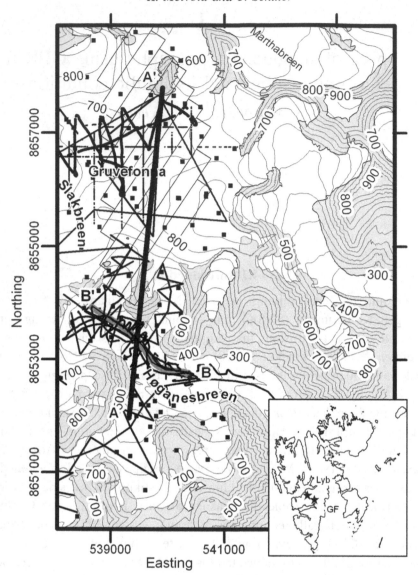

Figure 14.1.   Map of the Gruvefonna region based on SNSK 1:10 000 map of Svea Nord (50 m equidistance). The regions outside the glacier are shown in grey. Black lines show GPR tracks while black dots show borehole drill sites. Location of section A–A′ and location of GPR profile B–B′ shown in Figure 14.2a and b respectively are indicated.

therefore necessary to simulate the melt-water production at the glacier surface and to determine catchment areas from which water flows into the mine for different stages of mine enlargement. Here we report work on mapping ice thickness and subglacial topography and using these data to derive the actual

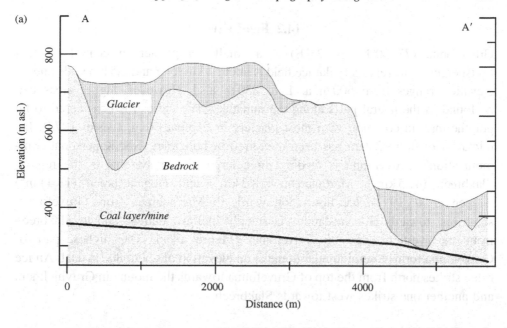

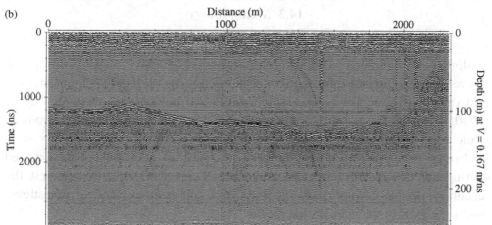

Figure 14.2. (a) Section A–A′ showing the mine situated in bedrock beneath the glacier bed. (b) A typical GPR profile running along the centre line of Høganesbreen from B to B′. For location of section and profile see Figure 14.1.

catchment area (Melvold *et al.* 2002, 2003, Schuler and Melvold 2004, Schuler *et al.* 2005). From these data, we calculate the theoretical distribution of hydraulic potential at the glacier bed and derive the probable catchment areas for different stages (panel-wise) of mine enlargement. The results of determining melt-water production on the glacier surface are reported in detail by Schuler and Melvold (2004) and Schuler *et al.* (2005).

14.2 Field site

Gruvefonna (77° 30′ N, 16° 25′ E) is a small ice glacier in central western Spitsbergen. This relatively flat ice field is about 3 km long and 2.5 km wide and its elevation ranges from 600 m a.s.l. to 880 m a.s.l. However, higher areas can be found in the lateral parts along the nunataks. The ice cap is considered to be polythermal, in common with most glaciers in Svalbard (e.g. Liestøl 1977). Ice thickness of up to 250 m has been measured in boreholes (SNSK personal communication). Gruvefonna is feeding three large glacier systems in the region: Slakbreen (41.5 km^2), Marthabreen (18.3 km^2) and Høganesbreen (13.4 km^2) (Hagen *et al.* 1993). Ice flows northwards to Marthabreen from Gruvefonna, southwest towards Høganesbreen via steeply inclined outlets, and to Slakbreen from the western edge of Gruvefonna (Figure 14.1). The highest part of Gruvefonna forms a semicircular dome at an elevation of about 800 m a.s.l. An ice ridge strikes north from the top of Gruvefonna towards the mountain Gruvhjelmen, and another one strikes west towards Slakbreen.

14.3 Methodology

14.3.1 Surface elevation data

A digital elevation model (DEM) of the surface topography was derived from the Store Norske Spitsbergen Kullkompani (©SNSK) 1:10 000 map entitled Gruvefonna. The map was generated using aerial photographs taken in August 1990. A 50 m resolution DEM was produced from this source using methods of Hutchinson (1989).

In addition we also used GPS-derived surface elevation data that were acquired during the GPR surveys in 2001 and 2004. These data were used to test the accuracy of the 1990 surface elevation data (on bedrock) and to investigate surface elevation changes on the glacier.

14.3.2 Ice thickness

In order to create maps of ice thickness and bed elevation of the glacier, both surface topography and a large number of ice thickness measurements are needed. The ice thickness was mapped using ground-penetrating radar (GPR) data and data from explorative drilling by the mining company.

Explorative drilling

Between 1986 and 1993, SNSK drilled several boreholes on the Gruvefonna glacier complex to map the extension of the coal layer in the bedrock beneath the glacier. While drilling, the thicknesses of both the ice and the bottom moraine (or debris-rich ice) were measured. The positions of drill sites were determined using traditional triangulation at relatively high accuracy (Figure 14.1).

GPR survey methods

GPR data were collected on the glacier complex between 1989 and 2005, providing wide spatial coverage (Figure 14.1). A typical profile of GPR sounding is illustrated in Figure 14.2b with the two-way traveltime of the electromagnetic wave on the *y*-axis and the traversed-over-snow distance along the surface on the *x*-axis. We picked (digitised) the time of the first detectable energy in the arrivals and the bed reflector from the raw data using Reflexw software (Sandmeier Scientific Software). The results of this processing are traveltimes for the direct and reflected waves on each radar trace. From the time difference between the two signals we determined the local ice thickness assuming a homogeneous propagation velocity of the electromagnetic wave in ice ($0.168 \, \text{m ns}^{-1}$) and simple parallel-planar geometry of the ice surface and bed in the vicinity of the measurement. Data migration was not carried out since the topography of the bedrock underneath Gruvefonna is simple and steep slopes are not observed (see Chapter 4). However, on some outlet glaciers, close to steep slopes, migration would have improved our ice thickness measurement, but since the main focus was the drainage of water underneath Gruvefonna no migration was carried out.

GPR data sets

(i)   The first data set (Hamran data) was collected in April and May 1989 by Hamran and Aarholt (1993), using a range-gated synthetic-pulse radar system with resistively loaded dipolar 5–20 MHz antennae. Sampling of the radar was triggered by a distance wheel at the rear of the sledge, giving equidistant sampling at an interval of 4.3 m. During this survey, navigation of traverses across the ice was done by compass bearing and the distance derived from the distance wheel of the sounder. Start and stop positioning on the glacier surface was obtained from Norsk Polarinstitutt maps (1:100 000) and it is assumed that the profile follows a straight line in between. The accuracy in position of these sounding lines is estimated to be within ± 200 m.

(ii) The second data set (UNIS-data) was collected in spring 2000, 2001 and 2002 by UNIS field courses with Christian Jaedicke and Martin Grønnevet as responsible scientists. The radar used was a pulse-radar system (pulseEKKO 100 from Sensors & Software Inc.). Either the 50 or the 100 MHz antenna was used. Towed by a snowmobile, the sampling of the radar was equidistant, triggered by a distance wheel at the rear of the sledge. A handheld GPS was used for navigation of all traverses on the glacier and the positions of the beginning and end of each were recorded. It is assumed that the profiles followed straight lines between start and stop positions and that the sampling was equidistant. The error due to the use of GPS is less than 10 m. The final position of the sounding line and individual traces is expected to be accurate within ± 25 m.

(iii) The third data set (UiO data) was collected in autumn 2001 by Kjetil Melvold and in spring 2005 by Thomas V. Schuler using a pulse-radar system (RAMAC/GPR from MALÅ Geoscience) and a low-frequeucy radav owned by Norsk Polarinstitutt, respectively . A 50 MHz and a 6.5 MHz antenna were used, respectively. Towed by a snowmobile, the sampling of the radar was performed at regular time intervals, resulting in different distance spatial sampling. The GPR transects were positioned by post-processed kinematic differential GPS measurements. The GPS data were acquired simultaneously with the GPR survey but at a different time interval. Scripts to match radar data and positions were developed, the output of which is ice thickness (H) and positioning co-ordinate (x,y,z) for the reflected wave for each radar trace. The accuracy of the GPS location along the profile is better than 0.15 m in horizontal and vertical positions.

14.3.3 Error analysis

The collection of data sets consists of 65 000 points (Figure 14.1), but the individual data sets are of varying coverage, scale and accuracy. Where sources data overlap, we have either to fuse or to select the source data. If the overlapping data sources have comparable accuracy and are complementary to each other, we integrate all the source data at the input level. If one data source is absolutely superior to others in terms of accuracy and density, we select the better data source and discard the others.

The uncertainties in the calculation of ice thickness are due to uncertainties in propagation velocity of the electromagnetic wave in ice, inaccuracies when picking reflectors (inaccurate traveltime determination), and the resolution of the radar system. The uncertainties of the velocity in the snow and ice are about $\pm 1\%$. This corresponds to an error of up to ± 5 m. The picking of the ice-base reflections has similar accuracy, and an error of ± 0.05 μs in traveltime corresponds to an error of ± 8 m in the ice thickness. The maximum resolutions that can be achieved correspond to 1/4 of the used wavelength, and thus range from c. 12.5 m to c. 0.5 m for the different wavelengths. In areas with steeper slopes the effect of not carrying out migration could significantly increase the error in estimated ice thickness (Moran et al. 2000). Larger errors in ice thickness could therefore be expected in such areas (see discussion).

In addition to uncertainties in ice thickness, the bedrock elevation is also affected by inaccurate positioning of the radar traces. To obtain a large data set for comparison and to account for errors that are associated with assumptions concerning the position of the GPR and ice thickness, we compared depth data from the different radar surveys at crossover points that were collected within a horizontal distance less than 30 m from each other. We obtained a set of 337 data points for validation. At each crossover point only the two nearest points from two different radar profiles were compared.

The maximum difference between the two data sets is 39 m (ice thickness was about 200 m) and the minimum difference is 0 m. On average, the Hamran data set contributes most to the deviation. The mean difference of crossover points with the Hamran data is 10 m, whereas the mean difference without Hamran data is only 6.8 m. In areas where differences are larger than 20 m, often in areas with steeper slopes, the less reliable data set has been moved to fit more reliable data or completely removed (e.g. Hamran data are less reliable than the UNIS data set due to less precise navigation). After this procedure all data sets were considered to be comparable and were combined to produce a thickness DEM as described below. The overall error in ice thickness was determined by considering the properties of the radar system and the accuracy of navigation. As a conservative number, we estimate this error to be ±20 m in the central study area. Figure 14.1 shows the distribution of the selected source data used in the final ice thickness DEM.

14.3.4 Data interpolation

A map of ice thickness was compiled using data from GPR, SNSK drilling and existing geodetic maps of the areas surrounding the glaciers. The outlines of the glacier were taken from the SNSK map. Some adjustments were made to these lines because moraine-covered ice had been interpreted previously as bedrock.

The ice thickness map was interpolated by applying different standard DEM modelling tools available in a geographic information system (GIS, ARC/INFO). The performance of different interpolation algorithms is strongly dependent on the pattern, density and format of source data (e.g. Liu *et al.* 1999). As such, the digital ice thickness map is based on data that are densely sampled along the radar profile but widely spaced between the lines (anisotropic distribution). On each sounding line the record represents a moving average of the real bed profile on a strip beneath the line. The width of the strip is typically of the order of 30 m, and the spacing between the lines is 0–1000 m. This pattern imposes serious difficulties on most general-purpose interpolation algorithms. We therefore used a procedure developed by Liu *et al.* (1999) using a combination of inverse distance weighting (IDW) and triangulated irregular network (TIN) methods. Smoothed contour maps were drawn from the first ice thickness map. The smoothed contour map was checked and contours redrawn manually, where necessary, before the final DEM and map were compiled.

The bedrock topography was obtained by subtracting the ice thickness from the surface elevation.

14.3.5 Theoretical drainage pattern

We used the spatial analysis capabilities of the GIS to assess the basic hydrological characteristics present beneath Gruvefonna. The procedure follows closely that described by Etzelmüller and Bjørnsson (2000) and Hagen et al. (2000).

In principle, the direction of water flow follows the gradient in hydraulic potential. At the bed of a glacier, the hydraulic potential is determined by both the topography of the glacier bed and the ice thickness of the overlying ice (Shreve 1972).

The hydraulic potential H, expressed as the height of a water column, is given by

$$H = z + h, \tag{14.1}$$

where z is the topographical potential of the glacier bed and h the hydraulic head. On the long-term scale, the value of h varies between 0 and the ice-flotation level h^* such that Equation (14.1) can be written

$$H = z + kh^*. \tag{14.2}$$

The factor k indicates the flow conditions, and we considered two situations: $k = 1$ and $k = 0$, characteristic limits of possible pressure variations. In the first case, $k = 1$, it is assumed that the water pressure is everywhere equal to the ice-overburden pressure. In this case, the direction of water flow will mainly be controlled by the surface gradient (surface topography of the glacier). In the second case, $k = 0$, water flows in a non-pressurized, open channel and the water flow follows the topographic gradient of the bed.

By mapping the hydraulic potential surface (Equation (14.2)) and assuming that the water flow is always directed down the maximum potential gradient (Shreve 1972), the possible location of subglacial flow paths can be derived. For our purpose, we used the contribution area approach in which, for each individual grid cell, the total is built of all grid cells in the upstream direction that drain to the considered grid cell. Based on the resulting potential map, the local flow direction is determined by adaptive filtering techniques, namely the D8 algorithm (O'Callaghan and Mark 1984). This algorithm determines the direction of the flow and quantifies the area that contributes to the considered grid cell. Preferential drainage pathways can be defined as those areas for which the contributing area exceeds a defined threshold.

In a glaciological context, D8 might be useful for representing a network of discrete conduits, although it still suffers from limited flow-orientation possibilities. Due to this we chose also to use a multiple-direction method which incorporates weighted area transfer, where all downstream neighbours among the eight nearest receive area in proportion to relative gradients.

Based on the distribution of the hydraulic potential and the upstream area distribution one can estimate the subglacial catchment area for different stages of mine enlargement for both cases $k = 0$ and $k = 1$.

14.4 Results

14.4.1 Ice thickness and bedrock topography beneath the glaciers

Figure 14.3 shows the ice thickness of the Gruvefonna glacier complex and Figure 14.4 shows the bedrock topography underneath the Gruvefonna glacier complex. The deeply incised valleys that lie beneath the centrelines of Slakbreen and Høganesbreen are dominant features. Beneath the eastern part of Gruvefonna, a mountain rises up. Beneath the western ice ridge from Gruvefonna, the topography is relatively flat. Ice thickness in this area is about 60–70 m. Beneath the western part of Gruvefonna, a deeply incised valley exists. This valley (called the Gruvefonna Valley) declines toward Slakbreen (southwesterly direction). The ice thickness in this valley is up to about 270 m. Maximum ice thickness of about 290 m was found beneath Slakbreen just downstream of the confluence of the Gruvefonna Valley and Slakbreen Valley. An overdeepened trough at this location indicates enhanced erosion in the confluence area. One can further see that the small tributary glaciers to the north and south of Høganesbreen are located in hanging valleys to the main Høganesbreen valley.

14.4.2 Hydraulic potential map

Contours of hydraulic potential assuming $k = 1$ are shown in Figure 14.5. Contours of potential for $k = 0$ are the same as bed elevation contours, which are shown in Figure 14.4. The distribution of hydraulic potential for $k = 1$ is somewhat parallel to contours of surface elevation, Figure 14.1.

14.4.3 Drainage pattern and drainage area

The potential map has been used to delineate the major subglacial glacier catchments shown in Figure 14.6 for $k = 1$. For comparison the surface drainage area is also shown (thick grey lines). Superimposed on Figure 14.6 are the main waterways beneath the Gruvefonna glacier complex for $k = 1$ predicted using the

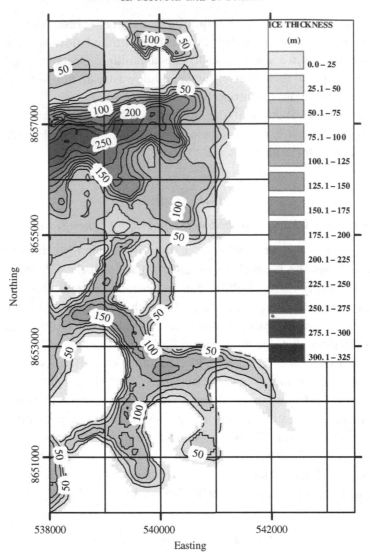

Figure 14.3.   Contour map of ice thickness (contour lines for every 25 m). The
map was constructed by interpolating extensive GPR sounding, borehole and
map data. The regions outside the glacier are shown in white.

D8 algorithm. Figure 14.6 also shows the upstream area distribution governed by
hydraulic potential for $k = 1$ using the multiple-direction method. Plots are presented as log10 of upstream area to remove the overwhelming downslope trend.
In Figure 14.6, dark grey indicates high values of upstream area, and thus
hydraulically preferential drainage pathways. Light grey areas have little to none
upstream area and therefore mark local water divides. Water is channelled
beneath the ice along the major valleys and tributaries. Most importantly, it is the

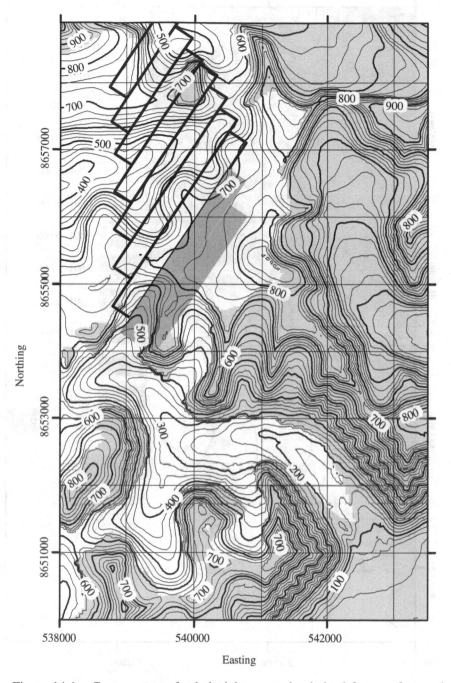

Figure 14.4.   Contour map of subglacial topography derived from surface and ice thickness maps. The regions outside the glacier are shown in grey. The area of the planned mine is outlined with bold lines.

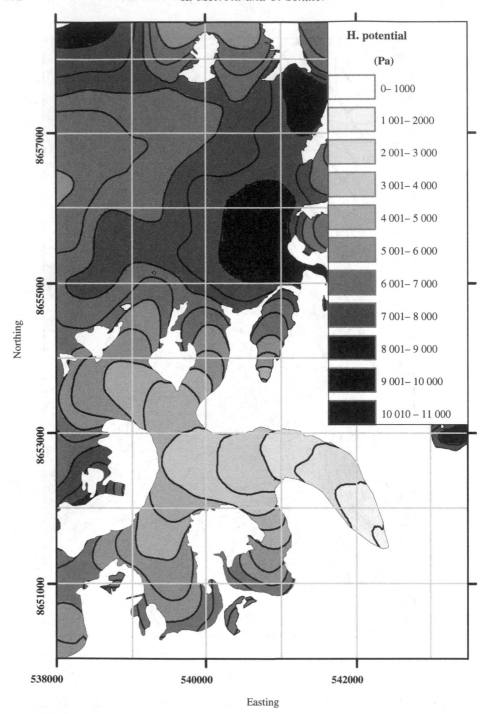

Figure 14.5. Contours of hydraulic potential (in Pa). The regions outside the
glacier are shown in white.

ice-surface slope that is sufficient to drive water downglacier. It is therefore unlikely that extensive water storage takes place along the glacier bed.

For $k = 1$, the subglacial catchment area for the first two panels of the mine is approximately 1.7 times larger than the mining area itself. If $k = 0$, the drainage area is approximately 1.8 times larger.

14.5 Discussion

Some remarks on the bedrock topography map follow. Along the glacier margins, the bedrock contours are incorrect due to the scale that has been used in the DEM production. If the ice thickness changes quickly towards the margin, only an average value will be represented in our DEM. In some areas this leads to large changes in elevation between neighbouring grid points and therefore more closely spaced contour lines than in reality (artificial cliffs along the glacier margin). Examples of this can be found along the northern margin of Høganesbreen.

Navigation errors (and therefore inaccurate positioning of the radar data) will have significant effects on the derived bedrock topography and the effect will depend on the surface slope. In steeper areas navigation errors will contribute more to the error in estimated bedrock topography than in less steep areas because ice thickness changes more rapidly in steep areas. Consequently, the less accurately positioned data from the earliest survey (Hamran data) were not used in areas with steep surface topography.

In addition, errors may occur in areas with steep valley walls since the data were not migrated. Moran *et al.* (2000) found that interpretation of raw-data profiles underestimated depth by 36% and that 2D migration underestimated depths by 16% compared to 3D migration in such areas.

Our map of ice thickness and bedrock topography will therefore not have homogeneousness accuracy over the whole domain. Thus, the maps may be expected to represent trends in the landscape in the steeper part of the glacier, in contrast to the less steep part of Gruvefonna. Furthermore, small-scale features will be preferably detected along the sounding lines. This is also true for the products derived from the ice thickness and bedrock topography maps, such as hydraulic potential, drainage paths etc. shown in Figures 14.5 and 14.6.

The hydraulic potential and upstream area distribution maps represent static situations and do not take in account any temporal or spatial pressure variations in the glacier hydrological system. The model implies that the transfer of water from the glacier surface to the mine always occurs along the same path with a hydraulic potential H somewhere between $k = 0$ and $k = 1$. This might not be the case as the mine is enlarged into an area where bedrock cover between the glacier bed and coal layer is significantly thinner ($<100\,\mathrm{m}$ in contrast to $>300\,\mathrm{m}$

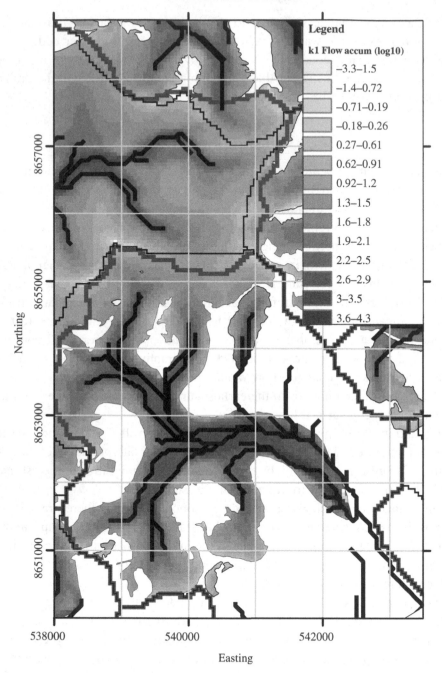

Figure 14.6.   Upstream area distribution governed by hydraulic potential for
$k = 1$. Plots are presented as log10 of upstream area to remove the dominantly
downslope trend. The main waterways beneath Gruvefonna (thick black lines),
subglacial catchments (thin black lines) and surface catchment areas (thick grey
lines). The regions outside the glacier are shown in white.

presently). Also the model does not take into account the fact that channels of different pressures will converge towards the one with the lowest pressure.

14.6 Conclusions

We have used an extensive GPR data set to assess the bedrock topography below the Gruvefonna glacier complex. The bed elevation DEM and surface DEM of the glacier were used to identify the theoretical structure of the subglacial drainage system for two cases: one where subglacial water pressure is high and one for water pressures equal to atmospheric pressure. From this distribution, the presumable area contributing to the mine was calculated with a difference between the above two cases of 10%. For water at high pressure, the mine captures from an area 1.7 times the area of the mine itself.

Acknowledgments

The authors wish to thank Y. Gjessing, C. Jaedicke and M. Grønnevet for their permission to use the radar data collected on the glacier complex during the UNIS course AGF-212 (Processes in ice and snow). Special thanks go to Trine Abrahamsen for her assistance in the field and her help with all the map data, and to Jørgen M. Stenvold. Sindre Vaagland provided energetic help during the fieldwork.

REFERENCES

Etzelmüller, B. and Bjørnsson, H. (2000). Map analysis techniques for glaciological applications. *International Journal Geographical Information Science*, **14**(6), 567–581.

Hagen, J. O., Liestøl, O., Roland, E. and Jørgensen, T. (1993). Glacier atlas of Svalbard and Jan Mayen. *Norsk Polarinstitutt Meddelelser*, **129**, 141pp. + maps.

Hagen, J. O., Etzelmüller, B. and Nuttall, A.-M. (2000). Runoff and drainage pattern derived from digital elevation models, Finsterwalderbreen, Svalbard. *Annals of Glaciology*, **31**, 147–152.

Hamran, S.-E. and Aarholt, E. (1993). Glacier study using wavenumber domain synthetic aperture radar. *Radio Science*, **28**(4), 559–570.

Hutchinson, M. F. (1989). A new procedure of gridding elevation and stream line data with automatic removal of spurious pits. *Journal of Hydrology*, **106**, 211–232.

Liestøl, O. (1977). Pingos, springs, and permafrost in Spitsbergen. *Norsk Polarinstitutt Årbok 1975*, 7–29.

Liu, H., Jezek, K. C. and Li, B. (1999). Development of an Antarctic digital elevation model by integrating cartographic and remotely sensed data: A geographic information system based approach. *Journal of Geophysical Research*, **104**(B10), 23199–23213.

Melvold, K., Vaksdal, M., Lappegard, G., Schuler, T. and Hagen, J. O. (2002).
   Potensiell drenering av vann fra Svea Nord gruva under Høganesbreen.
   Oppdragsrapport, Geografisk Institutt, Universitetet i Oslo, 30pp.
Melvold, K., Schuler, T. and Lappegard, G. (2003). Groundwater intrusions into a
   mine beneath Høganesbreen, Svalbard: assessing the possibility of evacuating
   water subglacially. *Annals of Glaciology*, **37**, 269–274.
Moran, M. L., Greenfield, R. J., Arcone, S. A. and Delaney, A. J. (2000). Delineation of a
   complexly dipping temperate glacier bed using short-pulse radar arrays. *Journal of
   Glaciology*, **46**(153), 274–286.
O'Callaghan, J. F. and Mark, D. M. (1984). The extraction of drainage networks
   from digital elevation data. *Computer Vision, Graphics and Image Processing*,
   28, 323–344.
Schuler, T. and Melvold, K. (2004). Melt water production at Gruvefonna and subglacial
   water intrusions in Svea Nord Gruva. *Oppdragsrapport*, Institutt for Geofag
   Institutt, Universitetet i Oslo, 50pp.
Schuler, T., Melvold, K., Hagen, J. O. and Hock, R. (2005). Assessing the future
   evolution of meltwater intrusions into a mine below Gruvefonna, Svalbard. *Annals
   of Glaciology*, **42**, 262–268.
Shreve, R. L. (1972). Movement of water in glaciers. *Journal of Glaciology*,
   11(62), 205–214.
Tolstikhin, N. I. and Tolstikhin, O. N. (1976). *Groundwater and Surface Water
   in the Permafrost Region*. Ottawa, Ontario, Inland Waters Directorate Department
   of the Environment, IWD Technical Bulletin No. 97. Translated from
   Chapter 9 of *Obshchee Merzlotovedenie (General Permafrost Studies)*, eds.
   Melnikov, P. I. and Tolstikhin, O. N., USSR Academy of Science, Siberian Branch,
   Novosibirsk.

15

Snow measurements using GPR: example from Amundsenisen, Svalbard

K. Melvold

15.1 Introduction

Snow distribution in mid- and high-latitude landscapes plays a key role in defining energy and moisture relationships associated with earth's climate system. The spatial distribution of snow is critical for accurate assessment and forecasting of snowmelt timing (Luce *et al.* 1998), snowmelt volume (Elder *et al.* 1991), avalanche hazard (Birkeland *et al.* 1995, Conway and Abrahamson 1984) and glacier mass balance modelling (e.g. Schuler *et al.* 2007) for initialisation of synoptic- and global-scale weather and climate models (Liston 1999). Traditional areal snow surveys are typically performed along lines, where snow depth measurements are done with graduated rods and snow density samples taken with a snow tube. Those methods are very time consuming and laborious and thus costly. Over the past decades, there has been a growing interest in looking at and developing new less time-consuming methods. Pomeroy and Gray (1995) gives a description of several of these methods.

The ground penetrating radar (GPR) technique for snow measurements has been developed over the past two or three decades (Annan *et al.* 1994, Ellerbruch and Boyne 1980, Kohler *et al.* 1997), but only recently have operational routines been established for using GPR in snow surveying (e.g. Marchand *et al.* 2001, Sand and Bruland 1998) and for snow distribution mapping on Svalbard glaciers (Sand *et al.* 2003, Taurisano *et al.* 2007, Winther *et al.* 1998). Snow depth is measured indirectly using GPR systems, also called georadar or snowradar since electromagnetic (micro) waves can easily penetrate ice and snow. Based on the traveltime of the radar pulse between the transmitter antenna, the ground and the receiver antenna, the snow depth can be calculated. Since the propagation velocity and thus traveltime varies with density and liquid water content in the snow, manual measurements are often used for calibration of the radar. Primarily, GPR

Applied Geophysics in Periglacial Environments, eds. C. Hauck and C. Kneisel. Published by Cambridge University Press. © Cambridge University Press 2008.

has yielded good results when the snow is 'dry' i.e. when there is little or no liquid water in the snow pack. Liquid water absorbs the microwaves, reducing the signal strength and thereby increasing measurement errors.

In this case study it will be shown how a ground-based GPR system coupled with a global positioning system (GPS) could be used for snow distribution mapping on the Amundsenisen ice cap, Svalbard. The main aim of this study has been to determine the near end-of-winter distribution of snow accumulation on Amundsenisen for the winter 2000/01. Snow accumulation patterns are a prerequisite to (i) determining surface elevation changes of the ice cap (Hagen *et al.* 2005); (ii) determining the mass balance; (iii) determining the water balance and (iv) sensitivity modelling of ice caps by means of energy-balance models. Furthermore, it is assumed that the near end-of-winter snow accumulation also gives a good indication of the regional distribution of winter precipitation. Hence, the study will also retrieve information about spatial distribution of precipitation on Amundsenisen.

Amundsenisen is a large accumulation plateau of $40\,km^2$ in southern Spitsbergen (Figure 15.1). This elongated (12 km long and 2–4 km width) relatively flat ice field lies mostly in the range 650–720 m a.s.l.; however, higher areas can be found in the lateral parts along the nunataks. Ice thickness of 580 m has been measured in a Russian borehole (Kotlyakov 1985). The ice cap is considered to be polythermal, in common with most glaciers in Svalbard (e.g. Liestøl 1977). Ground-based GPS measurements have shown a lowering of the surface elevation of the ice cap between 1991 and 2001 (Hagen *et al.* 2005). Amundsenisen is feeding three large glacier systems in the region: westwards to Høgestbreen, and further down Vestre Torellbreen ($338\,km^2$); southwestwards to Austre Torellbreen ($150\,km^2$); and southwards through Nornebreen and to Pairelbreen ($112\,km^2$) (Hagen *et al.* 1993) (Figure 15.1).

15.2 GPR and GPS equipment and measurements

GPR and GPS profiles were collected on Amundsenisen, at the end of April 2001, near end-of-winter when the snow pack is believed to be close to maximum thickness before the onset of melting. We used a RAMAC/GPR (MALÅ Geoscience, Malå, Sweden) equipped with a 500 MHz shield antenna and two JAVAD GPS receivers. Our GPR control unit, storage unit (a laptop computer), antennae and GPS system were mounted on a lightweight sledge pulled by a snowmobile. The radar antenna was mounted 2 m behind the sledge with the GPS and GPR units. The GPR traces were collected in the form of 1024 samples with a sampling frequency of 5889 MHz giving a time window of 179 ns for each trace. Eight individual radar returns were stacked to increase the signal-to-noise ratio. This stacked trace was recorded in the laptop computer. The radar was operated

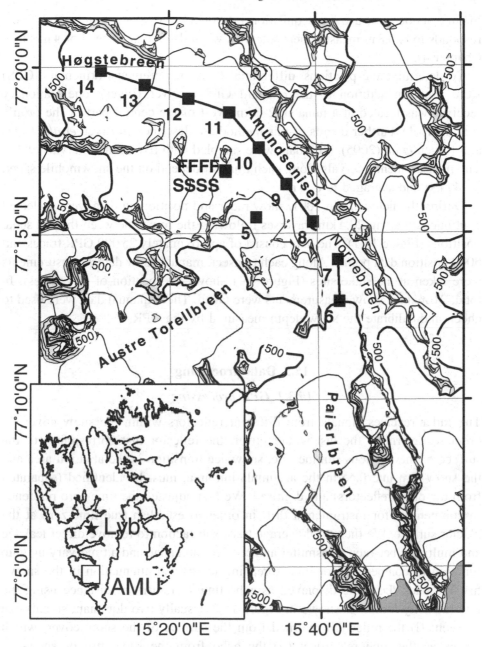

Figure 15.1. Map of the Amundsenisen ice cap, Svalbard, with a contour interval of 100 m. The glacierised, non-glacierised and sea areas are shown in white, light grey and dark grey respectively. Inset: map showing the Svalbard archipelago, indicating the location of the enlarged map. The squares mark the old stake positions. The GPS reference stations are marked with squares and labelled FFFF and SSSS. The location of the GPR profiles are shown by black lines.

in time mode with a trace acquisition rate of 5 traces per second, which corresponds to *c*. 0.8 m measurement intervals when driving at a speed of 4 m/s (see Chapter 4).

To determine the positions, differential GPS measurements (kinematic GPS) were used. In addition to the onboard GPS receiver (rover), one stationary receiver was located at a nunatak 2–3 km west of the central part of the profile (Figure 15.1). Further details of this method are described in Eiken *et al.* (1997) and Hagen *et al.* (2005). GPS data were sampled with fixed time intervals of 1 s. The measurement intervals along each transect depend on the snowmobile speed (*c*. 4 m/s) and averaged to *c*. 4 m.

During the fieldwork in 2001, snow radar and positioning data were collected continuously along 20.4 km traverses following the routes between the old stake positions (Figure 15.1). The data consist of approximately 25 000 GPR traces and 5000 position data points. Along each transect, manual snow depth measurements were taken at the stake sites (Figure 15.1 shows the location of stake sites). In total, 21 manual snow measurements were taken. These manual data were used to check and calibrate the snow depth measured by the GPR.

15.3 Data processing

15.3.1 GPR processing

The radar receives signals from different reflectors within the snow cover. In order to determine the last winter cover, the reflector that corresponds to the surface of the previous summer, i.e. snow–ice transition in the ablation area and the snow–firn interface in the accumulation area, must be identified (separated from the other reflectors) and digitised. We first adjusted the time-zero reference to compensate for instrumental drift in order to establish true time-zero at the glacier surface. We then used average trace subtraction to remove direct leakage and multiples between transmitter and receiver antennae and a time-varying gain function to compensate for both spreading losses and attenuation in the snow–firn–ice pack. Data are displayed as echo-time-of-return vs. distance using the line-intensity format as shown in Figure 15.2. Usually two dominant signals can be seen: (i) the reflection returned from the surface of the snow cover, which serves as the time reference; (ii) the echo from the snow–firn or snow–ice interface. The snow–ice/snow–firn transitions are revealed on the radar profile by the first continuous reflector, whereas occasional ice layers within the snow pack appear as shorter lower-amplitude reflectors. The digitisation was performed both by an automatic picking routine and manually, giving more or less equal results. However, at some places in the accumulation area, the snow–firn interface is less

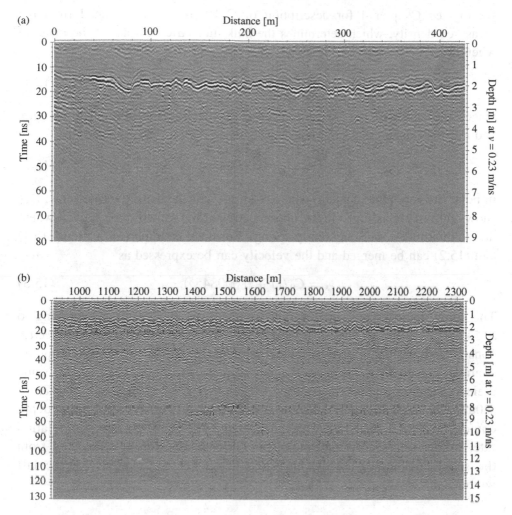

Figure 15.2.   (a) Processed radargram from the area close to stake S8 which shows a clear reflection created at the ice–snow interface. The distance between the traces is approximately 0.8 m and a two-way traveltime of 10 ns gives a depth of about 1.15 m in snow. (b) Radargrams obtained close to stake S6.

distinct, due to heterogeneous dielectric properties between the snow and the underlying firn, and a number of other factors described in Pinglot *et al.* (2001). In these cases, the last winter snow cover was identified using the layer reflection intensity correlated to the manual snow depth soundings.

In order to obtain quantitative snow accumulation measurements from the GPR images, one needs to convert time-dependent radar return signals to a depth. The depth of the reflector was calculated from the two-way traveltime and propagation velocity. The propagation velocity was not measured using the CMP set-up (see Chapter 4 for description of CMP) but was estimated from the measured density, which determines the bulk dielectric constant of the material when dry. The velocity (v_s) is given by

$$v_s = c_0/(\varepsilon)^{1/2}, \tag{15.1}$$

where c_0 (2.988×10^8 m s^{-1}) is the speed of the electromagnetic wave in vacuum, and ε is the relative dielectric constant. Kovacs *et al.* (1995) showed that

$$\varepsilon = (1 + 0.000845\delta)^2 \tag{15.2}$$

in polar firn (no liquid content) where δ (kg m^{-3}) is the density of the snow. Dry snow and firn are assumed in the upper metres of the snow/firn pack everywhere due to the time of the study (early spring). For these conditions Equations (15.1) and (15.2) can be merged and the velocity can be expressed as

$$v_s = C_0/(1 + 0.000845\delta). \tag{15.3}$$

The snow densities measured varied from 340 to 390 kg m^{-3}, averaging to 375 kg m^{-3}. Based on empirical models for estimation of the relative dielectric constant, this should give a dielectric constant close to 1.7, yielding an average velocity of 230 m μs^{-1} using Equation (15.3). This is the same dielectric constant and velocity used by Winther *et al.* (1998) and Sand *et al.* (2003) when investigating the snow distribution on Svalbard. The snow depth could thus be calculated from the measured TWT. Further, the accumulation rate (metre water equivalent) along the GPR profiles can be calculated from the snow thickness using the measured depth–density relations for the last winter snow cover.

15.3.2 Geo-coding

The geo-coding (to resolve the longitude and latitude position and elevation of the GPR trace) could be determined from the kinematic GPS survey since data from both systems were collected at fixed time increments (both GPR trace and GPS positions were time tagged). The minimum accuracy of the traces along the profile between each GPS position is estimated to be within 0.5 m both vertically and horizontally, but significantly better for most of the profile.

15.4 Results and discussion

In the processed radargrams (Figure 15.2a) the snow–ice interface (at a depth of approximately 2 m/18 ns) can clearly be seen. Below the snow–ice interface no

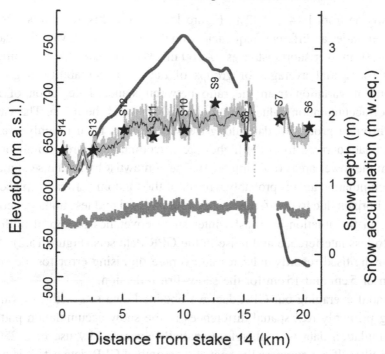

Figure 15.3. Snow thickness profile (light grey line), and the corresponding snow accumulation profile in metres water equivalents (m w.eq.) (grey line). Surface elevation in April 2001 is marked by the dark grey line. Both un-smoothed and smoothed (thin black line) GPR data are shown. Snow depth at old stake positions (labelled) is shown by black stars.

continuous reflector could be seen. Figure 15.2b shows a similar clear transition between snow and firn. Below this transition internal layers can be seen. Internal layers show large slopes along the profile and the vertical distance between these layers increases along the profile. Similar patterns have been reported previously on Svalbard (e.g. Pälli *et al.* 2002). Closely spaced layers indicate areas with less accumulation, whereas more widely spaced layers are associated with regions of higher accumulation. The increased separation thus indicates increased accumulation along the profile. Similarly, the thickness of the last winter snow layer increases along this profile.

In Figure 15.3 we present snow thickness profiles derived from the radar investigations complemented with surface elevations and manually measured snow thicknesses (probing) at the stake locations along the transect. From Figure 15.3 it is evident that there is an agreement between GPR and manual probing data at the stake sites. The maximum deviation was 15% or 0.25 m. In 2001 the snow cover was only 1.79 m thick (accumulation of 0.61 m w. eq.) on average with a standard deviation of 0.24 m; however, relatively large spatial variations exist along the profile ranging from 1.14 to 2.42 m (Figure 15.3). From this figure it is evident that snow cover varies at different frequencies and that peak to peak variation can easily reach 40–50 cm over short distances (0–100 m). We estimated the variability along transects for spatial averages of groups of data. The variability is given as a coefficient of variation from the ratio between standard deviation of a given number of accumulation values (300 points/200 m) and the mean. The variability for the different groups of data along transects ranges from roughly 8% for the groups of data derived from the higher area (accumulation area) to about 4% for data from the lower areas (ablation area). The somewhat higher noise signal found in the accumulation area is probably a result of the random vertical displacement of the GPR horizon due to interference of the radar signal and less homogeneous GPR horizons. In the ablation area last winter snow cover lies on top of uniform ice, leading to less interference and noise in the GPR data sets (Figure 15.2). Analysis of all our digitised records indicates an average digitising error for the snow–ice reflection of 5 cm and 15 cm for the snow–firn reflection.

By spatial averaging our GPR data we obtained data sets with low variability, reflecting primarily real spatial differences in the snow accumulation pattern. All the accumulation data were therefore spatially averaged by use of a 250 points Gaussian filter. If we compare the spatially smoothed GPR data sets it is clear that there are some deviations between the point measurements and the overall pattern (e.g. stake S9 and S11 in Figure 15.3). The smoothed GPR data give a better impression of the overall winter snow accumulation on the ice cap than the results from manual probing. From this it becomes clear that GPR sounding data yield extra information on winter snow accumulation distribution between the stake sites.

Closer inspection of the spatial distribution (along the transects) of the snow thickness data reveals a correlation between elevation and snow depth in the northern part of the transect (along Vestre Torellbreen) with values ranging from less than 0.5 m w.eq. at about 600 m a.s.l. to more than 0.65 m w.eq. at 750 m a.s.l. Along the central part of the ice cap (distance 9–15 km) the large-scale accumulation is relatively constant. The data show also that snow thickness is higher in the southern part (stakes S7 and S6) at similar heights than in the northern part of the transect. The amount of accumulation along the southern part of the transect is about 0.20 m higher than for the northern part and the accumulation gradient as a function of elevation is different between the two transects, respectively $\sim$0.10 m/100 m and $\sim$0.03 m/100 m. These two profiles are situated north and south of the main ice divide and show that there are different snow accumulation gradients on the northern and southern part of the ice cap.

At a distance of about 15 km a significant peak and trough occurs in the snow depth. These accumulation variations are located in a place where the topography changes from the relatively flat Amundsenisen to the steep and crevassed Nornebreen, indicating increased sourcing at the rim of Nornebreen and increased accumulation in the upper part of the crevassed area of Nornebreen. The lower part could not be surveyed due to heavy crevasses, hence there is a gap in the profile in Figure 15.3.

15.5 Conclusions

Comparison of the Amundsenisen winter mass-balance results from GPR surveys and manual probing shows that the first significant and continuous reflection detected in the GPR data corresponds to the snow–firn (above the firn line) and the snow–ice (below the firn line) interfaces. GPR data could be used for winter snow surveys and give extra information that cannot be obtained from normal manual snow probing. However, reliable measurements of snow thickness are only obtained when GPR is calibrated with manual depth measurements.

The winter snow accumulation is higher on the southern than on the northern side of the ice cap indicating that more winter accumulation comes with southerly than with northerly winds.

Acknowledgment

This study was financed by European Union project ICEMASS, through contract ENV4-CT97-0490. The work was conducted when the author was employed on a Postdoctoral Fellowship in the Department of Geosciences, University of Oslo under the supervision of Jon Ove Hagen. Logistics support by the Norwegian Polar Institute was greatly appreciated. Special thanks go to Jon Ove Hagen and Trond Eiken for help and support.

REFERENCES

Annan, A. P., Cosway, S. W. and Sigurdsson, T. (1994). GPR for snow pack water content. *Proceedings of the 5th International Conference on Ground Penetrating Radar*, Waterloo, Ontario, Canada.

Birkeland, K. W., Hansen, K. J. and Brown, R. L. (1995). Spatial variability of snow resistance on potential avalanche slopes. *Journal of Glaciology*, **41**, 183–189.

Conway, H. and Abrahamson, J. (1984). Snow stability index. *Journal of Glaciology*, **30**, 321–327.

Eiken, T., Hagen, J. O. and Melvold, K. (1997). Kinematic GPS survey of geometry changes on Svalbard glaciers. *Annals of Glaciology*, **24**, 157–163.

Elder, K., Dozier, J. and Michaelsen, J. (1991). Snow accumulation and distribution in an alpine watershed. *Water Resources Research*, **27**, 1541–1552.

Ellerbruch, D. A. and Boyne, H. S. (1980). Snow stratigraphy and water equivalence measured with active microwave system. *Journal of Glaciology*, **26**(94), 225–233.

Hagen, J. O., Liestøl, O., Roland, E. and Jørgensen, T. (1993). Glacier atlas of Svalbard and Jan Mayen. *Norsk Polarinstitutt Meddelelser*, **129**, Oslo, Norsk Polarinstitutt.

Hagen, J. O., Eiken, T., Kohler, J. and Melvold, K. (2005). Geometry changes on Svalbard glaciers: mass-balance or dynamic response? *Annals of Glaciology*, **42**, 255–261.

Kohler, J., Moore, J., Kennett, M., Engeset, R. and Elvehøy, H. (1997). Using ground-penetrating radar to image previous years' summer surface for mass-balance measurements. *Annals of Glaciology*, **24**, 355–360.

Kotlyakov, V. M. (1985). *Glatsyologia Spitsbergen*. Moscow, Nauka.

Kovacs, A., Gow, A. J. and Morey, R. M. (1995). The in-situ dielectric-constant of polar firn revisited. *Cold Regions Science and Technology*, **23**(3), 245–256.

Liestøl, O. (1977). Pingos, springs, and permafrost in Spitsbergen. *Norsk Polarinstitutt Årbok 1975*, 7–29.

Liston, G. E. (1999). Interrelationships among snow distribution, snowmelt, and snow cover depletion: Implications for atmospheric, hydrologic, and ecologic modeling. *Journal of Applied Meteorology*, **38**, 1474–1487.

Luce, C. H., Tarboton, D. G. and Cooly, R. R. (1998). The influence of the spatial distribution of snow on basin-averaged snowmelt. *Journal of Hydrology*, **12**(10–11), 1671–1683.

Marchand, W.-D., Bruland, O. and Killingtveit, Å. (2001). Improved measurements and analysis of spatial snow cover by combining a ground based radar system with a differential global positioning system receiver. *Nordic Hydrology*, **32**(3), 181–194.

Pälli, A., Kohler, J. C., Isaksson, E., Moore, J. C., Pinglot, J. F., Pohjola, V. A. and Samuelsson, H. (2002). Spatial and temporal variability of snow accumulation using ground-penetrating radar and ice cores on a Svalbard glacier. *Journal of Glaciology*, **48**(162), 417–424.

Pinglot, J. F., Hagen, J. O., Melvold, K., Eiken, T. and Vincent, C. (2001). A mean net accumulation pattern derived from radioactive layers and radar sounding on Austfonna, Nordaustlandet, Svalbard. *Journal of Glaciology*, **147**(159), 555–566.

Pomeroy, J. W. and Gray, D. M. (1995). *Snowcover Accumulation, Relocation and Management*. Saskatoon, Canada, National Hydrology Research Institute Science Report 7, Environment Canada.

Sand, K. and Bruland, O. (1998). Application of georadar for snow cover surveying. *Nordic Hydrology*, **29**(4/5), 361–370.

Sand, K., Winther, J. G., Maréchal, D., Bruland, O. and Melvold, K. (2003). Regional variations of snow accumulation on Spitsbergen, Svalbard, 1997–99. *Nordic Hydrology*, **34**(1/2), 17–32.

Schuler, V. T., Loe, E., Taurisano, A., Eiken, T., Hagen, J. O. and Kohler, J. (2007). Calibrating a surface mass-balance model for Austfonna ice cap, Svalbard. *Annals of Glaciology*, **46**, 241–248.

Taurisano, A., Schuler, V. T., Hagen, J. O., Eiken, T., Loe, E., Melvold, K. and Kohler, J. (2007). The distribution of snow accumulation across the Austfonna ice cap, Svalbard: direct measurements and modelling. *Polar Research*, **26**(1), 7–13.

Winther, J. G., Bruland, O., Sand, K., Killingtveit, A. and Marechal, D. (1998). Snow accumulation distribution on Spitsbergen, Svalbard, in 1997. *Polar Research*, **17**(2), 155–164.

16

Mapping frazil ice conditions in rivers using ground-penetrating radar

I. Berthling, H. Benjaminsen and A. Kvambekk

16.1 Introduction

In many high-latitude rivers, frazil ice production along open reaches sometimes causes blocking of the river course further downstream, beneath the
seasonal ice cover, and may eventually lead to flooding and formation of
extensive icing on the flood plain (Asvall 1998). When mitigation measures
must be taken, rapid mapping of the ice conditions beneath the surface ice
is advantageous. Therefore, the Norwegian Water Resources and Energy
Directorate have tested the use of ground-penetrating radar (GPR) for this
purpose.

The dielectric contrasts between ice and both water and wet sediments are
large and clear reflections are expected (Table 16.1). Beneath the surface ice
cover, a mixture of water and a variable amount of ice crystals (frazil ice) may be
found. It is likely that this mixture of ice and water will cause clutter (chaotic
returns from material inhomogeneity) and influence the velocity of a passing
electromagnetic signal.

Airborne radar technology has been successfully applied to measure ice
thickness on rivers and lakes (e.g. Arcone and Delaney 1987, Arcone 1991,
Leconte and Klassen 1991, Arcone *et al.* 1997), and ground-based surveys of ice
thickness have for instance been standardised by Sensors & Software Inc. through
designated 'ice picker' software for use with their Noggin 500 MHz system.
Other aspects of river ice, such as frazil ice and bottom ice, have received less
attention, although there is a study by Dean (1977) and detection of frazil ice by
airborne radar is also mentioned by Steven Arcone on the CRREL website (www.
crrel.usace.army.mil/sid/gpr/Airborne_GPR.html).

Applied Geophysics in Periglacial Environments, eds. C. Hauck and C. Kneisel. Published by Cambridge
University Press. © Cambridge University Press 2008.

Table 16.1. *Typical values for relevant parameters*

Medium	Dielectric constant (ε_r)	Typical velocity (m/ns)	Vertical resolution (m)
Ice	3–4	0.16	0.35
Saturated sand	20–30	0.06	0.15
Water	80	0.033	0.08

From Davis and Annan (1989).
Vertical resolution is set to $\lambda/2$, where $\lambda \approx c/\left(f(\varepsilon_r)^{1/2}\right)$.
Here, c is the velocity of light and f is the centre frequency of 200 MHz.

16.2 Setting and field procedures

An upstream tributary to the River Trysil in the southeastern part of Norway is mainly not ice covered during winter, due to warm water outflow from a lake. In response to anticyclonic conditions in wintertime, the area experiences cold ambient temperatures, often dropping beneath $-20\,°C$. Such circumstances lead to potentially large frazil ice production along this open reach, and this regularly causes blocking of the Trysil along a shallow reach further downstream. GPR profiles were collected in a cross-section along this shallow reach.

Fieldwork for the study was carried out on 10–11 January 2002. We used a pulseEKKO 100 radar system from Sensors & Software Inc. with a fast port, equipped with antennae of 100 and 200 MHz centre frequency. The 200 MHz antennae proved to yield the most interesting results, and only these data are presented here. The 200 MHz transmitter and receiver antennae were fixed at 0.5 metre spacing, with antennae orientated transverse to the profile direction. The radar was run in 'continuous' mode with a delay between each shot set equivalent to a step size of about 0.1 m for the estimated towing speed. The time window was set to 303 ns, resulting in a depth window of 5 m at a medium velocity of 0.033 m/ns, and the sampling interval was 800 ps. Based on initial tests, one profile was selected for more detailed investigations. A marker was placed every 5 m along this profile before GPR data were collected. During GPR data collection, a tick-mark was introduced on the radar profile when passing each of these markers. After GPR data collection, the ice was drilled through at each of these tick-marked positions, and 'ground truth' information on surface ice thickness, density and thickness of frazil ice and depth to the river bottom was collected. The frazil ice data were collected by measuring the resistance to penetration of a probing rod. Therefore, only qualitative information on density is available. This ground truth information made it possible to compare subsurface conditions for specific traces along the GPR profile.

During the survey, the radar console, 12V batteries and the computer were carried by an operator, while the antennae with their electronics and batteries were mounted on a sledge, towed by another person. Although a standard set-up, this system proved very vulnerable. Ambient temperatures below $-20\,°C$ made connecting all components troublesome, and cables and connections seemed to become more fragile. Consequently, an improved system set-up was developed, where a computer with external display (with an optional heater) and keyboard was used. Then, radar console, computer and external 12V batteries for both console and computer could be mounted within a storage box on the sledge (see Figure 4.2, Chapter 4). The system is controlled by the operator from the external display and keyboard, which is connected to the computer through a rugged cable. The improvements include less-fragile connections and better protection from low ambient temperatures, as well as a smaller total number of cable connections in the field, thus reducing total survey time. The modified system could be operated by one person, but an additional person to tow the sledge along the profile lines is preferable. The new set-up potentially induces more noise (see Chapter 4), and has only briefly been tested on river ice conditions. Initial tests showed that the set-up works, but that the quality of data was not as good as the original. Comparative tests on lake ice also showed that noise is an issue (Melvold, personal communication 2006), so that the use of two sledges generally must be recommended. In that case, it may still be possible to utilise the advantages of the set-up described above, although the overall ease of handling is reduced.

16.3 Results

Figure 16.1 shows the ground-penetrating radar profile across the river along with the locations of boreholes. In Figure 16.2, data on ice thickness, frazil ice conditions and depth to the riverbed are illustrated for the 11 boreholes. One interesting feature of the GPR profile is that the returns from the reflector corresponding to the riverbed seem to be dislocated in the vertical direction in some areas. These areas correspond fairly well to the parts of the profile where frazil ice is dense. From Table 16.1 it is seen that EM wave velocity is about five times larger in ice than in fresh water. One would assume a variable wave velocity in frazil ice (in response to variations in frazil ice density), but with values between those of ice and water. Thus, the presence of frazil ice will cause differences in two-way traveltime (TWTT) to the bottom reflector. To analyse this further we computed traveltimes for the frazil ice layer by considering the TWTT of the riverbed reflector to be the sum of the TWTTs of direct and ground

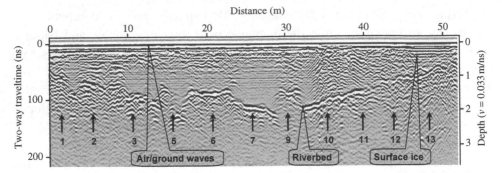

Figure 16.1. GPR profile across River Trysil with 200 MHz antennae. The only processing steps applied are the removal of redundant traces due to short stops during profiling, and a constant gain multiplier of 50. Numbers and arrows point to locations where boreholes through the ice were made and ground truth information collected after radar profiling. The velocity used for depth conversion in this profile is that of water (0.033 m/ns).

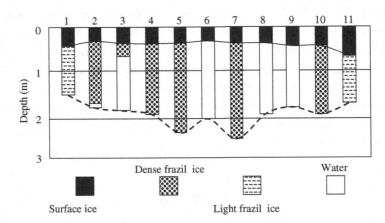

Figure 16.2. Measurements in boreholes, showing depth to the river bed (total length of the rectangles), frazil ice thickness and density (either 'dense' or 'light'), and thickness of surface ice cover. The solid line shows surface ice, the stippled line the approximate bottom topography.

waves and the waves through surface ice, frazil ice and water:

$$\text{TWTT}_{\text{frazil}} = \text{TWTT}_{\text{riverbed reflector}} - \text{TWTT}_{\text{direct/ground wave}}$$
$$- \text{TWTT}_{\text{ice}} - \text{TWTT}_{\text{water}}. \quad (16.1)$$

First we selected the trace nearest to the corresponding boreholes. As examples, the traces closest to boreholes 6 and 7 are displayed in Figure 16.3. From these traces, TWTT to the riverbed reflector and TWTT of the direct and ground waves was determined, while TWTT of surface ice and water were computed based on

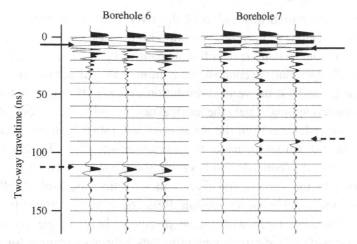

Figure 16.3.   GPR traces corresponding to boreholes 6 and 7. The vertical axis displays TWTT. In borehole 6, no frazil ice was found and mainly the direct and ground waves are present besides the river bottom reflector. Solid arrows point to the ground wave and dotted arrows to the river bottom reflector. The signal from the bottom reflector arrives much earlier in borehole 7, even though the riverbed here is about 0.5 m deeper than in borehole 6 (Figure 16.2).

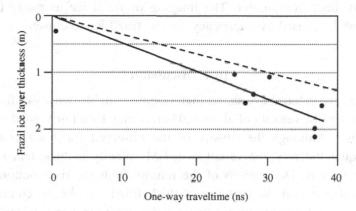

Figure 16.4.   Plot of calculated traveltime in the frazil ice layer against measured thickness of the same layer. The regression line is forced through zero to reveal an average EM velocity of frazil ice. The dotted line shows the EM velocity of water.

measured thicknesses (Figure 16.2) and standard velocities (Table 16.1). This analysis was done for each borehole where frazil ice was observed. Frazil ice traveltimes and thicknesses were finally plotted against each other, and a regression line (forced through zero) was computed. This gave a mean EM velocity of frazil ice of 0.049 m/ns, with a degree of explanation of 0.77 (Figure 16.4). A comparison with the velocity for water shows that the points closest to the velocity of water are the two boreholes with 'light' frazil ice.

Apart from the partial dislocation of the river bottom reflector, the bottom is easily recognised along this profile. This was also the case during profiling, although the presence of boulders on the river bottom causes this interface sometimes to look like a series of hyperbolas. In contrast, the bottom of the surface ice cover is difficult to see. From Table 16.1, the vertical resolution in ice is about 0.35 m for 200 MHz antennae, which is approximately the ice thickness found on the investigated profile. Thus, only one reflection from the ice is obtained, rather than two clearly distinguishable reflections from the upper and the lower interfaces. Thicker ice close to one of the riverbanks (borehole 11) is visible though (Figure 16.1).

In addition to the surface and river bottom reflectors, further reflectors are found within the water/frazil ice section. Reflections also originate from beneath the river bottom, but these will not be discussed here. In the water/frazil ice section, some parts of the profile generally lack reflectors (except for some faint 'ringing' seen as horizontal lines across the profile), while chaotic or hyperbolic reflectors are found elsewhere. Sometimes these reflectors seem to be clearly linked to the frazil ice (boreholes 1, 2, 7, 10), but there are boreholes with frazil ice that show these reflectors only in the upper parts (boreholes 4, 5). On the other hand, boreholes 8 and 9 show very clear reflectors from within the water column, without any frazil ice present. The imaging of frazil ice as partly hyperbolic features may be caused by a tendency for the frazil ice to cluster.

16.4 Discussion

The analysis presented above shows that, despite variable density of the frazil ice in question, an EM velocity of about 0.05 m/ns may be a fairly good assumption for frazil ice. Although the density of the observed frazil ice could not be described quantitatively, the variation in EM velocity is consistent with qualitative observations. Dislocations of the returns from the river bottom reflector may be used to reveal the presence of thick frazil ice. As the contrast in EM velocity is so much larger between water and surface ice than between water and frazil ice, however, relatively small variations in surface ice thickness could also cause such observations. Dislocations of the river bottom reflector can therefore not be used as the only criterion.

Reflectors were also present within the frazil ice/water column. Almost all locations where frazil ice is present in the boreholes show reflections more or less all the way between surface ice and river bottom. Combinations of dislocations and reflections may then be a reasonable way of detecting frazil ice occurrences. Nevertheless, further knowledge concerning the cause of reflections within a streaming water body, such as those visible in the area of boreholes 8 and 9, is necessary before firm conclusions can be drawn.

Concerning the application of GPR for frazil ice detection, one must distinguish between operative usage of GPR in situations where a channel through the ice must be constructed to prevent or reduce upstream flooding, and research on frazil ice accumulation as such. In the first case, GPR is only of interest if it can be applied quickly and the results are available in real-time. For short profiles, coring will be simpler and quicker. If larger areas must be investigated, GPR may be of value if the information can be obtained in real-time. In that respect, the example shown here is rather promising. If ice conditions are more complicated, such as layered ice with water inclusions or remnants of ice-runs, reliable information could be difficult to obtain. Furthermore, a flexible type of GPR system, such as the pulseEKKO 100, is not well suited for applications where 'quick-launch' and surveying speed are vital.

In the second case, the presented case study shows that frazil ice could be mapped with GPR, thus providing a method to distinguish zones where frazil ice tends to accumulate over large areas. If time for detailed post-processing is available, influences of more complex surface ice conditions can be reduced or removed. Furthermore, use of several antenna frequencies might add to the possible information one could gather from the GPR method.

16.5 Conclusions

This case study discussed the ability of a GPR system with 200 MHz antennae to locate frazil ice occurrences in rivers. Based on a comparison of GPR data and ground truth information we conclude that:

- The presence of frazil ice influenced the EM velocity along the trace.
- EM velocity of frazil ice was approximately 0.050 m/ns, with slightly lesser values when the frazil ice was very light.
- Zones of frazil ice were also visible from reflections in the column between river bottom and surface ice, although exceptions were found with respect to both frazil ice zones with fewer reflections and patches of streaming water with reflectors.

REFERENCES

Arcone, S. A. (1991). Dielectric-constant and layer-thickness interpretation of helicopter-borne short-pulse radar wave-forms reflected from wet and dry river-ice sheets. *IEEE Transactions on Geoscience and Remote Sensing*, **29**, 768–777.

Arcone, S. A. and Delaney, A. J. (1987). Airborne river ice thickness profiling with helicopter-borne UHF short-pulse radar. *Journal of Glaciology*, **33**, 330–340.

Arcone, S. A., Yankielun, N. E. and Chacho, E. F. (1997). Reflection profiling of arctic lake ice using microwave FM-CW radar. *IEEE Transactions on Geoscience and Remote Sensing*, **34**, 436–443.

Asvall, R. P. (1998). Frazil ice formation causing flooding. *Proceedings of the 14th International Symposium on Ice*, Potsdam, NY, USA, 663–667.

Davis, J. L. and Annan, A. P. (1989). Ground-penetrating radar for high-resolution mapping of soil and rock stratigraphy. *Geophysical Prospecting*, **37**, 531–551.

Dean, A. M. (1977). *Remote Sensing of Accumulated Frazil Ice and Brash Ice in the St. Lawrence River*. CRREL Report, 77–8.

Leconte, R. and Klassen, P. D. (1991). Lake and river ice investigations in Northern Manitoba using airborne SAR imagery. *Arctic*, **44**, 153–163.

Appendix

Tables of geophysical parameters for periglacial environments

Table A.1. *Characteristics of geophysical techniques*

Method	Applications (references point to the previous chapters – further applications are mentioned and cited in the respective chapters)	Persons needed for survey	Comments
Electrical resistivity tomography (ERT)	Detection of massive ice in rock glaciers, moraines and other periglacial phenomena (Chapters 1, 6, 9) Mapping isolated ice occurrences (Chapters 1, 6, 8, 10) Monitoring the temporal evolution of permafrost and visualising transient processes (Chapters 1, 6) Analysing the ice origin in rock glaciers (Chapter 6) Quantifying/comparing ice content (Chapters 1, 10)	1 or 2	• Obtaining good electrical contact between electrodes and ground is essential • Experience in data inversion is needed for data processing • Differentiation between ice, air and special rock types can sometimes be difficult • Insensitive to man-made EM noise
Vertical electrical soundings (VES)	Detection of significant ground ice in rock glaciers, glacier forefields and homogeneous permafrost regions (Chapters 5, 7, 8, 9) Classification of sounding curves in typologies for typical landforms in periglacial environments (Chapters 5, 9)	3 or 4	• Obtaining good electrical contact between electrodes and ground is essential • Lightweight compared to ERT • Assumption of lateral homogeneous ground conditions • Insensitive to man-made EM noise

Method	Applications		Characteristics
Frequency-domain electromagnetic induction (FEM) mapping	Mapping isolated ice occurrences (Chapter 2) Mapping the boundaries of periglacial phenomena (Chapter 2) Mapping horizontal differences in the active layer thickness (Chapters 2, 8) Determining the amount of heterogeneity to assess the representativeness of single point measurements (Chapters 2, 8)	1 or 2	• Lightweight equipment • Different surface conditions may have a large influence on the survey results • Instrument drift may lead to erroneous results due to small measurement values • Simple data processing • Results can be frequency-dependent • Sensitive to man-made EM noise (power lines etc.)
Time-domain electromagnetic induction (TEM) sounding	Determining the thickness of a permafrost layer (Chapters 2, 7)	1 or 2, 3 for carrying	• Often no resolution in uppermost 5–10 m • Poor in resolving the exact resistivity value of a resistive middle layer • Usually large penetration depths • Sensitive to man-made EM noise (power lines etc.)
Radio-magnetotelluric (RMT)	In principle similar applications as for ERT surveys are possible – as the method is comparatively novel only few applications in periglacial terrain exist so far (Chapter 2)	1 or 2	• Lightweight equipment • Frequency-dependent results • depends on the existence of a recordable transmitter

227

Table A.1. (*cont.*)

Method	Applications (references point to the previous chapters – further applications are mentioned and cited in the respective chapters)	Persons needed for survey	Comments
Refraction seismic tomography	Detection of massive ice in rock glaciers, moraines and other periglacial phenomena (Chapters 3, 9) Mapping isolated ice occurrences (Chapter 10) Differentiation between ice, air and special rock types, each exhibiting anomalously high resistivity values (Chapters 9, 10) Mapping the active layer thickness (Chapters 3, 9, 10)	2 or 3	• Number of receivers should be 12 at least, with shots between every or every second receiver location • Sledgehammer as source is sufficient for most applications • Geophones are sensitive to wind and rain, leading to noisy data sets • Experience in data inversion is needed for data processing
Ground-penetrating radar (GPR)	Delineation of the boundaries of massive ice in rock glaciers and other periglacial phenomena (Chapters 4, 12) Mapping the active layer thickness (Chapters 4, 12) Mapping of subglacial topography (Chapter 14) Monitoring of seasonal hydrothermal changes in glaciers (Chapter 13) Ice and snow thickness mapping (Chapters 15, 16) Delineation of thickness and layering of talus deposits (Chapter 11)	2–4, depending on the terrain	• Small penetration depth in the case of conductive near-surface layers • Difficult to apply in very heterogeneous media • Fast survey speeds on ice and snow • Choice of suitable antennae frequency often important • Difficulties in determining the depth scale of the radargram (velocity–depth conversion) • Experience in data processing is needed • Sensitive to man-made EM noise (power lines etc.) and reflectors above ground (rock wall, cables etc.)

228

Method	Measured property	Penetration depth	Data processing	Power requirements/logistics
ERT	Electrical resistivity	0.15–0.2 times current electrode spacing (Wenner array)	Software packages available (e.g. RES2DINV); comparatively easy	Power supply through rechargeable battery packs; the use of spare batteries is recommended
VES	Electrical resistivity	1/6–1/8 maximum electrode spacing	Simple algorithms and software are available for 1D-inversion	Small; rechargeable battery packs; very long cables are required – care has to be taken to avoid entangling cables on the drums
FEM mapping	Electrical conductivity	Depends on instrument geometry and frequency (skin depth, see Chapter 2); often restricted to <10 m	Often direct conductivity reading at the instrument	Small; often commercially available batteries or rechargeable data-loggers
TEM sounding	Electrical conductivity	Depends on upper-layer resistivity and loop size (diffusion depth, see Section 2.5)	Software packages available (e.g. Terraplus TEMIX), similar to VES	Power supply through rechargeable battery packs; the use of spare batteries is recommended

Table A.1. (*cont.*)

Method	Measured property	Penetration depth	Data processing	Power requirements/ logistics
RMT	Electrical resistivity and phase	Depends on measurement geometry and frequency (skin depth, see Chapter 2); usually larger than ERT	Simple plotting of horizontal profiles or 2D-inversion algorithms	Power supply for the receiver through rechargeable battery packs
Refraction seismic tomography	Seismic P-wave velocity	1/3–1/5 of the offset distance (max. shot–receiver distance); depends also on shot energy and velocity distribution	First arrival picking. Software packages available (e.g. REFLEXW); some experience needed	Rechargeable battery for the seismograph
GPR	Dielectric properties; EM wave velocity	Difficult to predict; depends on attenuation and frequency	Software packages available (e.g. REFLEXW); experience needed	Power supply through rechargeable battery packs; a laptop computer needed for surveying

Table A.2. *Compilation of resistivity values for various materials and different periglacial regions*

Material	Electrical resistivity (kΩ m)	Reference
Clay	0.001–0.1	Chapter 1
Sand	0.1–5	Chapter 1
Gravel	0.1–0.4	Chapter 1
Granite	5–1000	Chapter 1
Gneiss	0.1–1	Chapter 1
Schist	0.1–10	Chapter 1
Ground water	0.01–0.3	Chapter 1
Frozen sediments/ground ice/mountain permafrost	1–1000	Chapter 1
Glacier ice (temperate)	1000–100 000	Chapter 1
Air	infinity	Chapter 1

Material/location	Region	Electrical resistivity		Reference
		Permafrost (kΩ m)	Unfrozen layer (kΩ m)	
Murtel rock glacier	Swiss Alps	200–2000	10–30	Chapters 1, 9
Stelvio rock glacier	Italian Alps	~500	10–30	Chapter 1
Muragl rock glacier	Swiss Alps	100–1000	10–20	Chapters 1, 2
Rock glaciers (blocky) with frozen sediments	Swiss Alps & Pyrenees	50–500		Chapter 5
Kanchenjunga rock glacier	Eastern Nepal	>2000		Chapter 6
Bouldery rock glaciers	Swiss Alps	45–410	10–50	Chapter 9
Pebbly rock glaciers	Swiss Alps	0.5–20	1–5	Chapter 9
Vegetated (prob. relict) rock glaciers	Swiss Alps	0.5–10	1–5	Chapter 9
Rock glacier Büz North (pebbly)	Swiss Alps	3	2	Chapter 9

Table A.2. (cont.)

Material/location	Region	Electrical resistivity		Reference
		Permafrost (kΩ m)	Unfrozen layer (kΩ m)	
Debris-covered glaciers, buried ice, ice-cored moraines (proximal side)	Swiss Alps & Pyrenees	>1000		Chapters 1, 5
Kanchenjunga glacier	Eastern Nepal	>1000		Chapter 6
Ice-cored moraines (distal side), upper part of rock glaciers	Swiss Alps & Pyrenees	100–500		Chapters 1, 5
Gandegg ice-cored moraine	Swiss Alps	100–200	10–20	Chapter 1
Ice-cored moraine	Sweden	100–>1000	5–30	Chapter 1
Patterned ground with sorted polygons	Sweden	10–30	1–10	Chapter 1
Gelifluction slope	Sweden	50–500	10–30	Chapter 1
Ice-cored moraine	Iceland	100–500	5–20	Chapter 1
Solifluction slope	Iceland	10–50	1–10	Chapter 1
Palsa	Iceland	5–30	0.1–5	Chapter 1
Muragl glacier forefield: coarse mat.	Swiss Alps	100–500	5–20	Chapter 1
Muragl glacier forefield: fine mat.	Swiss Alps	20–100	5–20	Chapter 1
Glacier forefield with (possibly) degrading permafrost	Swiss Alps & Pyrenees	20–200		Chapters 1, 5

Glacier forefield with coarse debris	Swiss Alps & Pyrenees		10–50	Chapter 5
Schilthorn summit (frozen bedrock)	Swiss Alps	3–5	0.5–2	Chapter 2
Jotunheimen (frozen bedrock)	Norway	10–20	1–3	Chapter 2
Kasprowy Wierch summit area (bedrock)	Tatra Mountains, Poland	30–90	10–15	Chapter 8
Daisetsu Mountains (frozen ground)	Japan	50–100		Chapter 6
Moskuslagoon (silty clay)	Svalbard	0.04	0.001–0.01	Chapter 7
Caribou-Poker Creek (clayey silt, VES)	Central Alaska	1.1–7.3	0.6–0.7	Chapter 7
Caribou-Poker Creek (clayey silt, TEM)	Central Alaska	0.5–1.6		Chapter 7
Nalaikh (discontinuous permafrost)	Central Mongolia	0.063		Chapter 7
Khatgal (continuous permafrost)	Northern Mongolia	0.5		Chapter 7
Shijir valley (forested talus slope)	Northeastern Mongolia	0.5–1.6		Chapter 6
Val Bever low-altitude scree slope	Swiss Alps	50–150	10–40	Chapters 1, 10
Präg low-altitude scree slope	Black Forest, Germany	20–80	20–80	Chapter 10
Zastler low-altitude scree slope	Black Forest, Germany	10–200	10–80	Chapter 10
La Glacière low-altitude scree slope	Vosges, eastern France	50–130	20–100	Chapter 10

Table A.3. *Compilation of P-wave velocity values for various materials and different periglacial regions*

Material	P-wave velocity (m/s)	Reference
Air	330	Chapter 3
Water	1500	Chapter 3
Topsoil	100–600	Chapter 3
Peat	200–800	Chapter 3
Clay	550–2700	Chapter 3
Loam	200–1900	Chapter 3
Loess	250–1200	Chapter 3
Sand	200–2000	Chapter 3
Gravel	150–2000	Chapter 3
Sandstone	800–4500	Chapter 3
Marlstone	1300–4500	Chapter 3
Dolomite	2000–6200	Chapter 3
Limestone	2000–6200	Chapter 3
Magmatic rock	2400–5100	Chapter 3
Metamorphic rock	3000–5800	Chapter 3
Talus deposits	550–2500	Chapter 3
Till	1500–2700	Chapter 3
Glacial ice	3100–4500	Chapter 3
Permafrost	2400–4300	Chapter 3

Observational site	Region	P-wave velocity (m/s)		Reference
		Permafrost	Unfrozen layer	
Turtmann valley, talus slope	Swiss Alps	2500–4400	600–1500	Chapter 3
Bouldery rock glaciers	Swiss Alps	2100–4400	350–650	Chapter 9
Pebbly rock glaciers	Swiss Alps	2000–3200	330–450	Chapter 9
Vegetated (prob. relict) rock glaciers	Swiss Alps	1600–3100	360–950	Chapter 9
Rock glacier Murtèl (bouldery)	Swiss Alps	3700	350	Chapter 9
Rock glacier Büz North (pebbly)	Swiss Alps	2800	390	Chapter 9
Gelifluction slope	Sweden	2000–4500	500–1500	Chapter 1
Präg low-altitude scree slope	Black Forest, Germany	2000–2600	400–800	Chapter 10
Zastler low-altitude scree slope	Black Forest, Germany	1000–2000	400–800	Chapter 10
La Glacière low-altitude scree slope	Vosges, eastern France	1000–1300	330–700	Chapter 10
Val Bever low-altitude scree slope	Swiss Alps	2000–4500	600–1300	Chapters 1, 10

Table A.4. *Compilation of electrical permittivity and velocity values for various materials and different periglacial regions*

Material/location	Region	Electric permittivity (ε_r)	Velocity (m/ns)	Reference
Air		1	0.3	
Ice		3–4	0.168	Chapters 14, 16
Frazil ice (River Trysil)	Southeastern Norway		0.049	Chapter 16
Dry snow, polar firn		1.7	0.23	Chapter 15
Saturated sand		20–30	0.06	Chapter 16
Sediments		~25		Chapter 13
Water (fresh)		80	0.033	Chapters 4, 13, 16
Permafrost materials		4–8	0.11–0.15	Chapter 4
Typical active layer		25	0.06	Chapter 4
Limestone		4–8		Chapter 11
Dolostone		4–8		Chapter 11
Gravel			0.09	Chapter 11
Gravel, loose heap			0.16–0.18	Chapter 11
Talus sediments (average)			0.1–0.14	Chapter 11
Loose debris (dolostone), Parzinn	Austrian Alps	4.6–9	0.11–0.13	Chapter 11
Grass covered debris (dolostone), Tegelberg	German Alps		0.105	Chapter 11
Loose debris (limestone) Dammkar	German Alps		0.1	Chapter 11

Loose debris (limestone), Zugspitze	German Alps	0.12–0.128	Chapter 11
Very loose and fresh debris (limestone), Arnspitze	Austrian Alps	0.14	Chapter 11
Loose debris (gneiss, mica-schist), Kühtai	Austrian Alps	0.09–0.115	Chapter 11
Loose coarse debris (gneiss, mica-schist), Turtmanntal	Swiss Alps	0.12–0.14	Chapter 11
Loose debris (pyroclastic rocks), Cwm Cneifion	Snowdonia, Wales	0.14	Chapter 11
Grass-covered debris, Cwm Cneifion	Snowdonia, Wales	0.1	Chapter 11
Moraine (Egesen), Turtmanntal	Swiss Alps	0.105	Chapter 11
Moraine (LIA), Turtmanntal	Swiss Alps	0.095	Chapter 11
Moraine (Egesen), Parzinn	Austrian Alps	0.09	Chapter 11
Rock glacier, Prins Karls Forlandet	Svalbard	0.14	Chapter 12

Index

Printed in the United States
By Bookmasters